Conference Proceedings of the Society for Experimental Mechanics Series

Series Editor

Kristin B. Zimmerman
Society for Experimental Mechanics, Inc.,
Bethel, USA

The Conference Proceedings of the Society for Experimental Mechanics Series presents early findings and case studies from a wide range of fundamental and applied work across the broad range of fields that comprise Experimental Mechanics. Series volumes follow the principle tracks or focus topics featured in each of the Society's two annual conferences: IMAC, A Conference and Exposition on Structural Dynamics, and the Society's Annual Conference & Exposition and will address critical areas of interest to researchers and design engineers working in all areas of Structural Dynamics, Solid Mechanics and Materials Research.

Brandon Dilworth · Timothy Marinone · Jon Furlich
Editors

Topics in Modal Analysis & Parameter Identification II, Vol. 10

Proceedings of the 43rd IMAC, A Conference and Exposition on Structural Dynamics 2025

River Publishers

Published, sold and distributed by:

River Publishers
Broagervej 10
9260 Gistrup
Denmark

www.riverpublishers.com

ISBN 97887-438-0155-9 (Hardback)

ISBN 97887-438-0167-2 (eBook)

https://doi.org/10.13052/9788743801559

Conference Proceedings of the Society for Experimental Mechanics

An imprint of River Publishers

Preface

Topics in Modal Analysis & Parameter Identification II represents one of twelve volumes of technical papers presented at the 43rd IMAC, A Conference and Exposition on Structural Dynamics, organized by the Society for Experimental Mechanics, and held February 10-13, 2025. The full proceedings also include volumes on Nonlinear Structures & Systems; Dynamics of Civil Structures; Model Validation and Uncertainty Quantification; Dynamic Substructuring & Transfer Path Analysis; Special Topics in Structural Dynamics & Experimental Techniques; Computer Vision & Laser Vibrometry; Dynamic Environments Testing; Sensors & Instrumentation and Aircraft/Aerospace Testing Techniques; Data Science in Engineering; and Structural Health Monitoring & Machine Learning.

Each collection presents early findings from experimental and computational investigations on an important area within Structural Dynamics. Topics in Modal Analysis and Parameter Identification represents papers on enabling technologies for dynamic systems and applications of Parameter Identification in specific application areas.

The organizers would like to thank the authors, presenters, session organizers, and session chairs for their participation in this track.

Editor:
Brandon Dilworth–*MIT Lincoln Laboratory, MA, USA;* Timothy Marinone–*ATA Engineering, Inc., CA, USA;* Jon Furlich–*ATA Engineering, Inc., CA, USA.*

Contents

Chapter 1
Modal Identification from Dynamic Signals Obtained from OFDR Distributed Fiber Optics

Patrick Brewick

Abstract Among the many sensing technologies utilized for health monitoring, fiber optics are emerging as a popular choice for monitoring non-standard structures such as pipelines, offshore platforms, and marine structures due to their imperviousness to environmental disturbances and flexible sensing arrangements. Further, the current state-of-the-art in fiber optic technologies can embed dense arrays of sensing points along a single, continuous fiber, creating powerful distributed sensing systems. Optical frequency domain reflectometry (OFDR) is one such distributed sensing technique that utilizes weak continuous fiber Bragg gratings (FBGs) to infer strain along the length of the fiber. Recent advances in OFDR interrogation techniques allow for sampling rates fast enough to capture dynamic effects.

This study will present some results obtained from performing experimental modal analysis on a cantilevered plate using an OFDR distributed fiber optic sensing system (FOSS). Two different dynamic quantities of interest are produced by the OFDR FOSS, both the strains and carrier frequencies at each FBG. Modal analysis is performed using the blind source separation method on both sets of dynamic time histories, yielding high quality estimates of modal frequencies and mode shapes. Details on the quality of modal identification with carrier frequencies will be discussed and a comparison against the strain-based results will be presented.

Keywords Distributed fiber optics · modal identification · strain sensing · carrier frequencies · blind source separation

Introduction

Structural health monitoring (SHM) has been utilized across the engineering community to collect data from a variety of systems and structures, spanning civil infrastructure [1], aerospace structures [2] and marine craft [3], among other dynamical systems [4]. Amidst the different sensor technologies commonly deployed for SHM, fiber optic sensing systems (FOSS) have emerged as a powerful tool for conducting SHM via strain sensing [5–7], where FOSS constitute a broad class of sensors that use various optical phenomena, e.g., intensity modulation, scattering, optical loss, and polarization, to respond to strain [8].

The increasing utilization of FOSS for SHM solutions is partly due to its invulnerability to harsh conditions since the fiber is insensitive to electro-magnetic effects, resistant to corrosion, and maintains reliable operability across a wide range of temperatures [9]. The relative flexibility of fiber also allows it to be outfitted in a variety of geometric patterns and designs [10], matching other highly configurable and modular sensing technologies. These desirable qualities have made FOSS a popular choice for marine and aerospace applications, but efforts to monitor civil infrastructure have also benefited from FOSS [5]. One particular challenge that has arisen, however, is that SHM of large civil structures typically operates on vibrations-based monitoring [11], but the dynamic strains induced by operational loading conditions are often quite small, e.g., below one microstrain [12]. In the past, this limited the utilization of even traditional strain sensors for operational (OMA) to only a few studies, e.g., [13, 14].

Recent developments in the optical interrogation and signal processing of fiber-based strain sensors have demonstrated that sufficient sensitivity and accuracy are, in fact, attainable for vibrations-based SHM with FOSS [12]. Further, the high sensitivity of strain-based features to local damage [15] has led strain modes to emerge as an attractive alternative to more conventional displacement-based mode shapes [16]. For instance, utilizing a series of discrete fiber Bragg grating (FBG) sensors, Wang et al. [17] recovered strain mode shapes and modal parameters by fitting polynomials to the strain power

Patrick Brewick
Department of Civil and Environmental Engineering & Earth Sciences
University of Notre Dame, Notre Dame, IN 46561
e-mail: pbrewick@nd.edu

Brandon Dilworth et al. (eds.), *Topics in Modal Analysis & Parameter Identification II, Vol. 10*, Conference Proceedings of the Society for Experimental Mechanics Series, https://doi.org/10.13052/9788743801559-1

spectral density functions. Anastasopoulos et al. have similarly deployed discrete FBGs to collect strain information and subsequently perform OMA using stochastic subspace identification techniques on a variety of structures, including experimental prestressed concrete beams [12], a steel I-beam [16] and a steel arch bridge [18].

Distributed sensing, such as that offered by more advanced FOSS [19], provides a contrast to traditional SHM that uses discrete sensing points since distributed FOSS yields a high density of sampling points all within a single sensor (fiber). While several techniques have been employed for distributed sensing [20], NASA has developed a distributed FOSS that utilizes optical frequency domain reflectometry (OFDR) technology. OFDR has been shown to reliably provide distributed strain measurements over long gauge fibers [21], such as those several meters in length. For instance, an optical fiber that spans 40 feet can support FBGs at one-half-inch intervals, providing ≈1,000 strain sensing points along a single fiber [22]. OFDR utilizes amplitude modulation of a continuous-wave light source, such as a laser, to obtain backscatter information in the frequency domain. Then, by using specialized fiber that contain very weak FBGs, reflections along the fiber can be enhanced to allow strain information to be inferred from the shifts in reflection [21].

This study applies the principals of OFDR to perform experimental modal analysis on a cantilevered plate. Traditional OFDR takes the frequency domain information from the FOSS and transforms it into an impulse response in the time domain through an inverse fast Fourier transform (IFFT) [21]. Applied strain causes linear shifts in the resonant wavelengths of the impulse response, which is how FOSS monitors strain [23]. However, applied strains also lead to changes in the frequency domain, which register as Doppler shifts. This study considers both the strain estimates recovered from the IFFT while also leveraging the Doppler shift data from the frequency domain, performing modal analysis on both sets of time series data. The results from the Doppler shift time histories are subsequently compared to a traditional strain-based identification. This study demonstrates that valuable modal information can be conveyed by operating directly on the original frequency domain data.

Background

The FOSS developed by NASA is based on the Langley method, using OFDR to interrogate the FBGs [23]. The sensing arm is composed of single mode fiber with continuous weak FBGs. FBGs reflect light from a particular wavelength, commonly referred to as the Bragg wavelength λ_B, while transmitting light from other wavelengths. For this system, the FBGs are written in the same wavelength, i.e., every grating has the same λ_B.

The light reflected from each weak FBG is combined with a reference signal to create an interference signal, and the resulting intensity of this interfering signal is measured by the photodetector. When received by the interrogator within the data acquisition system, the output signal represents the summation of the intensity of the interfering signal from each FBG along the fiber length, producing the type of interferogram shown in Fig. 1a. Applying a fast Fourier transform (FFT) to a chosen interferogram maps the intensity information from the wavelength domain to the spatial (frequency) domain, as shown in Fig. 1b. The resulting spatial domain information shows a series of reflections that appear as "top hats," as highlighted in Fig. 1b. These top hats are the carrier frequencies, which represent each FBG at its unique distance (position) along the sensing fiber. The resonant wavelength of each FBG is found by isolating an individual top hat and taking the IFFT from the spatial domain back to the wavelength domain.

Applied strain has a linear relationship to shifts in resonant wavelength — straining an FBG in tension or compression causes it to stretch or contract, respectively, which produces an increase or decrease in the wavelength. Thus, monitoring changes in this wavelength allows for strain to be inferred at a given FBG [23]. Assuming conventional temperature correction techniques [24], the expression relating pure mechanical strain ε to wavelength shifts is given as [25]

$$\varepsilon = \frac{1}{1-p_e}\frac{\Delta\lambda_B}{\lambda_B} \tag{1}$$

where $\Delta\lambda_B$ is the change in the resonant (Bragg) wavelength and $1/(1-p_e)$ represents a strain-optic calibration coefficient. Evaluating this expression at every sample time for a given FBG location generates its strain time history. When subjected to an applied strain the top hats also experience a shift in their carrier frequency; however, these are Doppler shifts that accumulate along the length of the fiber [26]. The difference in the position of the center of mass of each top hat, i.e., the difference in the identified carrier frequency, is proportional to the integrated strain along the length of the fiber [26].

Strain measurements, including those collected by fiber optic strain sensors, can be decomposed into a modal basis [25], as shown in Eq. (2)

$$\boldsymbol{\varepsilon}(t) = \boldsymbol{\Psi}\boldsymbol{\xi}(t) \tag{2}$$

where $\boldsymbol{\varepsilon}(t)$ is a vector containing the measured strains, $\boldsymbol{\Psi}$ is the matrix collecting normalized strain relative to the mode shape, i.e., the strain mode shape matrix, and $\boldsymbol{\xi}(t)$ is the vector of modal weights, i.e., modal coordinate vector. Equation (2)

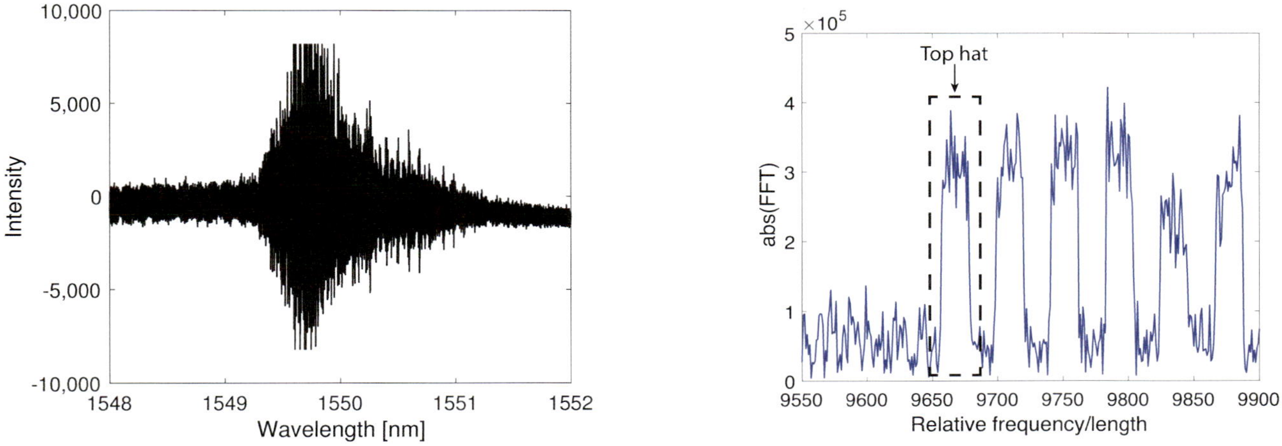

(a) Inteferogram

(b) FFT

Fig. 1 Original interferogram (wavelength domain) and corresponding FFT (spatial/frequency domain). Note that the FFT is shown over a windowed range for enhanced visualization.

can be utilized for modal identification, wherein modal information, such as mode shapes, are extracted from the original strain measurements. For instance, blind source separation (BSS) is a family of time domain-based techniques [27] for system identification that assumes a set of observations $\mathbf{x}$ are mixture of a set of independent sources $\mathbf{s}$, as shown in Eq. (3) with $\mathbf{A}$ denoting the "mixing" matrix.

$$\mathbf{x} = \mathbf{A}\mathbf{s}\,. \tag{3}$$

The aim of BSS is to use the observed outputs to solve for the original sources, as shown in Eq. (4)

$$\hat{\mathbf{s}} = \mathbf{W}\mathbf{x} \tag{4}$$

where $\hat{\mathbf{s}}$ is the vector of estimated source signals and $\mathbf{W}$ is the "de-mixing" matrix. It can be easily observed that Eq. (2) assumes the same form as Eq. (3), wherein the observed strains serve as the mixed signals, the modal coordinates are the independent sources, and the mode shape matrix is a mixing matrix. This study adopts the second order blind identification (SOBI) [28] to perform modal analysis and ultimately identify the strain mode shapes.

SOBI is also applied to the the time histories created by the change in carrier frequencies at a given sensing point. These carrier frequency shifts can be monitored at every sample time by tracking the change in the center of mass of each top hat, creating a time history of the Doppler shifts for a given sensing point or FBG. As mentioned previously, the shift in carrier frequency is proportional to the integrated strain, which is related to the slope of the displacement. Thus, applying the SOBI method to these time histories produces estimates of the slope mode shapes, as will be further demonstrated in the results.

Analysis

A series of tests were performed on a cantilevered test article at the NASA Neil A. Armstrong Flight Research Center. The test article is a 6061 aluminum plate 70 in. long (span) × 12 in. wide × 1/4 in. thick. The plate was cantilevered with one end bolted to a frame (clamped) while the other end remained free. The interrogating fiber for the FOSS starts 6 in. from the clamped end and includes a total of 81 FBGs along its span, as shown in the schematic illustration in Fig. 2a. A finite element (FE) model of the test article was created in FEMAP [29]. A modal analysis was performed on this model to identify the natural frequencies and mode shapes for the first four bending modes. Figure 2b shows the displacement mode shape of Mode 3 generated by the FE model. In addition to the displacement mode shapes, the strain mode shapes are also recovered from the FE model wherein profiles are extracted along the centerline of the top surface of the plate given that this is where the fiber is placed.

The test article was excited by a shaker attached to the tip of the free end that provided sinusoidal excitation at three designated frequencies. The three excitation frequencies were chosen based on the natural frequencies identified from the modal analysis of the FE model such that the test operating at 10 Hz was close to the 2nd natural frequency, 28 Hz was close to the 3rd natural frequency, and 55 Hz was close to the 4th natural frequency. Thus, each test targeted a particular

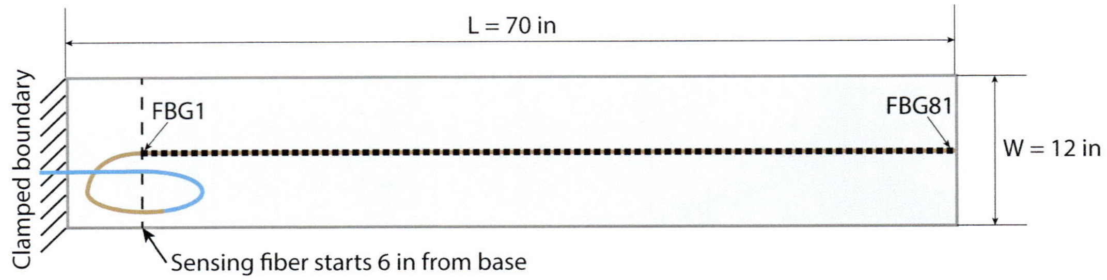

(a) Top view schematic of sensing fiber placement

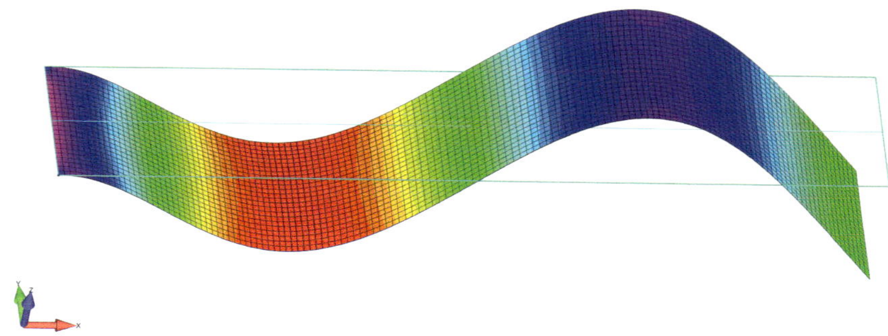

(b) Displacement mode shape of Mode 3

Fig. 2 Cantilevered wing test article showing a) illustrative schematic of sensing fiber placement and b) modal analysis of finite element model.

mode. Each test lasted 15-25 s with an average test length of about 21 s. During testing, the unit had an acquisition rate of approximately 120 Hz.

When computing the strain from fiber optic measurements using Eq. (1), the change in wavelength $\Delta\lambda_B$ is always relative to some baseline. Since each test features a quiescent period of 2-4 s at its start, the baseline for each test is calculated using its first 60 measurements, which translates to the first one-half second of each recording. Once computed, all 81 strain time histories are used within the SOBI method. The first vector of the estimated mixing matrix from the 10 Hz test is plotted against the corresponding strain mode shape from the FE model in Fig. 3. Note that the strain

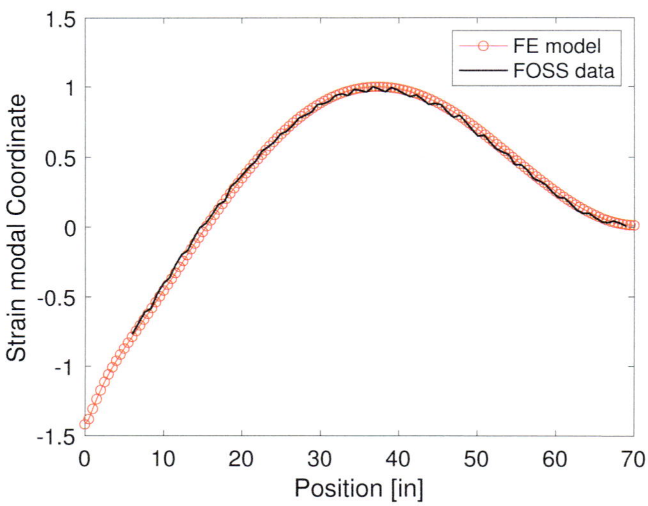

(a) Strain mode shape

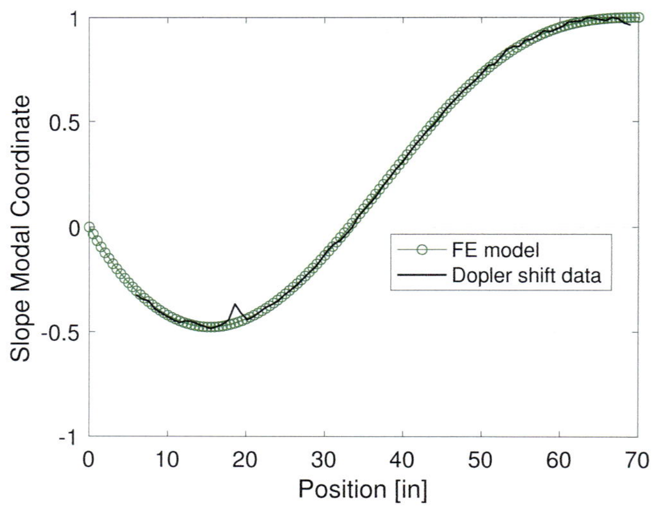

(b) Slope mode shape

Fig. 3 Comparison of a) strain mode shapes and b) slope mode shapes for Mode 2 taken from the FE model and estimated from SOBI using FOSS data from 10 Hz test.

mode shapes are normalized by their maximum value, which occurs around 37 in. for both mode shapes. Also, note that estimated strain mode shape begins at 6 in. because that is the location of the first FOSS sensing point, as previously detailed.

The Doppler shift time histories are analyzed in a similar fashion as the strain time histories. One notable difference is that the Doppler shifts are relative to the initial position, meaning the baseline for the Doppler shifts is simply the initial measurement. Recall that the Doppler shift is proportional to the change in (displacement) mode shape $\boldsymbol{\phi}$ along the fiber length x, i.e., $\frac{d\boldsymbol{\phi}}{dx}$. Thus, the identified mode shape is related to the "slope" of the displacement mode shape and has to be compared to the corresponding "slope" mode shape from the FE model. The slope mode shapes in the FE model can be found by either differentiating the displacement mode shape or by integrating the strain mode shape. This study computes the FE model-based slope mode shapes through integration, and Fig. 3b compares the model-based slope mode shape against the first mode shape identified from the Doppler shift-based mixing matrix from the 10 Hz test. As can be clearly observed, the identified slope mode shape is a strong match for the model-based slope mode shape.

The similarity between mode shapes can be quantified through the modal assurance criterion (MAC) [30], which measures the correlation between mode shapes. For the purposes of this study, the MAC value is defined as

$$\mathrm{MAC}(i,j) = \frac{\left|\mathbf{A}_i^{\mathrm{T}} \boldsymbol{\psi}_j^{\mathrm{FE}}\right|^2}{\left(\mathbf{A}_i^{\mathrm{T}} \mathbf{A}_i\right)\left(\boldsymbol{\psi}_j^{\mathrm{FE}^{\mathrm{T}}} \boldsymbol{\psi}_j^{\mathrm{FE}}\right)} \tag{5}$$

where $\boldsymbol{\psi}_j^{\mathrm{FE}}$ is the strain mode shape vector from the jth mode of the FE model and $\mathbf{A}_i$ is the ith column of the mixing matrix. The MAC value can only be computed from 6 to 70 in. due to the lack of sensing fiber along the first 6 in.; thus, the strain mode shapes are technically incomplete, which could slightly affect the computation of the associated MAC values. Note that Eq. (5) can also be used to compute the MAC values for the slope mode shapes with $\frac{d\boldsymbol{\phi}}{dx}^{\mathrm{FE}}$ serving as the reference mode shapes from the FE model.

Table 1 MAC value comparison for identified strain and slope mode shapes

Test Excitation Frequency	Strain Mode No.			Slope Mode No.		
[Hz]	2	3	4	2	3	4
10	**0.999**	**0.960**	0.359	**0.999**	**0.988**	0.782
28	0.241	**0.997**	0.206	**0.929**	**0.997**	0.723
55	0.234	0.293	**0.996**	**0.937**	0.743	**0.998**

The MAC analysis process is repeated for the strain and Doppler shift data from all three tests wherein the strain mode shapes from the FE model are compared against the identified strain mode shapes and the integrated strain mode shapes from the FE model, i.e., slope mode shapes, are compared against the identified slope mode shapes. The results are shown in Table 1. The targeted modes (diagonal entries) yield comparably high MAC values for both data sources, but the slope mode shapes have much higher MAC values for the non-targeted (off-diagonal) modes in every case. The 10 Hz test results in multiple modes with MAC values larger than 0.90 using both strain data and the Doppler shifts — MAC values greater than 0.90 are highlighted in bold — but this is the only such instance for the strain data. In contrast, the Doppler shift data yields two such MAC values for every test. This is significant because it suggests that the Doppler shift can reliably yield modal information beyond the single mode being targeted by the excitation.

Figure 4 presents two of the identified slope mode shapes for non-targeted modes. Figure 4a presents a highly compelling estimate of the slope mode shape for Mode 3 from the 10 Hz test. While the estimate for the slope mode shape for Mode 2 in Fig. 4b is much less convincing, it nonetheless correlates with the true slope mode shape, indicating that the Doppler shifts contain some modal energy from those non-targeted modes. Taken together, Table 1 and Figs. 3b and 4 demonstrate that the Doppler shift time histories contain richer modal information than strain, which suggests that dynamic signals generated from distributed FOSS systems based on OFDR possess high value modal information beyond their typical strain measurements.

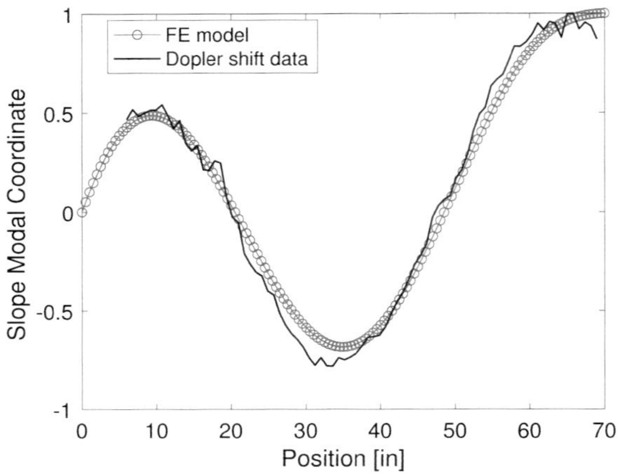

(a) Mode 3 – 10 Hz Test

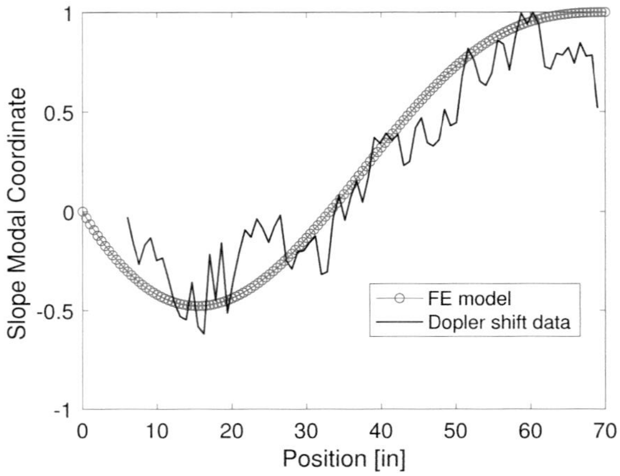

(b) Mode 2 – 28 Hz Test

Fig. 4 Comparison of non-targeted slope mode shapes taken from the FE model and estimated from SOBI using FOSS data.

Conclusion

This study shows that interferograms produced from OFDR-based distributed FOSS can yield both strain time histories as well as time histories of the Doppler shifts in carrier frequencies of each weak FBG sensor along the fiber. System identification via blind source separation can then be applied to these dynamical responses to recover modal information, such as mode shapes estimates. The proposed identification approach was applied to FOSS data collected on a cantilevered plate subjected to experimental modal analysis. Using the SOBI method, strain mode shapes are recovered from strain time histories while the Doppler shift time histories yield estimates of the slope mode shapes. Using MAC values based on comparisons to reference mode shapes from an FE model, it was shown that both dynamical responses produce high quality mode shape estimates for the modes targeted by the excitation.

Crucially, this study also revealed that the Doppler shift time histories contain richer modal information than strain. This was demonstrated by the Doppler shift information producing multiple high quality mode shape estimates, i.e., MAC values greater than 0.90, for every test, while the strain time histories had only managed this once. Further, the typical MAC value produced by the slope mode shapes was noticeably larger than those from the strain data. The results of this study demonstrate that OFDR-based distributed FOSS contain multiple dynamic signals that possess high-value modal information for SHM applications, and that Doppler shift time histories should be the subject of further investigation, e.g., under operational loading conditions.

Acknowledgments The author would like to acknowledge Dr. Allen Parker, Mr. Frank Peña, and Dr. Patrick Chan at the NASA Armstrong Flight Research Center for their generosity in sharing the data from the test article as well as their several helpful discussions.

References

1. Brownjohn, J. "Structural health monitoring of civil infrastructure". *Philosophical Transactions of the Royal Society A: Mathematical, Physical and Engineering Sciences*, 365(1851):589–622 (2007).
2. Diamanti, K. and Soutis, C. "Structural health monitoring techniques for aircraft composite structures". *Progress in Aerospace Sciences*, 46(8):342 – 352 (2010).
3. Chen, S., Wang, J., Zhang, C., Li, M., Li, N., Wu, H., Liu, Y., Peng, W., and Song, Y. "Marine structural health monitoring with optical fiber sensors: A review". *Sensors*, 23(4) (2023).
4. Worden, K., Farrar, C.R., Haywood, J., and Todd, M. "A review of nonlinear dynamics applications to structural health monitoring". *Structural Control and Health Monitoring*, 15(4):540–567 (2008).
5. Glisic, B. and Inaudi, D. *Fibre optic methods for structural health monitoring*. John Wiley & Sons (2008).
6. Dalla Vedova, M.D.L., Berri, P.C., Maggiore, P., and Quattrocchi, G. "Design and development of innovative FBG-based fiber optic sensors for aerospace applications". *Journal of Physics: Conference Series*, 1589(1):012012 (2020).
7. Min, R., Liu, Z., Pereira, L., Yang, C., Sui, Q., and Marques, C. "Optical fiber sensing for marine environment and marine structural health monitoring: A review". *Optics & Laser Technology*, 140:107082 (2021).

8. Udd, E. "An overview of fiber-optic sensors". *Review of Scientific Instruments*, 66(8):4015–4030 (1995).
9. Murayama, H., Kageyama, K., Naruse, H., Shimada, A., and Uzawa, K. "Application of fiber-optic distributed sensors to health monitoring for full-scale composite structures". *Journal of Intelligent Material Systems and Structures*, 14(1):3–13 (2003).
10. Cazzulani, G., Silva, A., and Pennacchi, P. "Optimization of continuous sensor placement for modal analysis: Application to an optical backscatter reflectometry strain sensor". *Mechanical Systems and Signal Processing*, 150:107242 (2021).
11. Brownjohn, J.M.W., De Stefano, A., Xu, Y.L., Wenzel, H., and Aktan, A.E. "Vibration-based monitoring of civil infrastructure: challenges and successes". *Journal of Civil Structural Health Monitoring*, 1(3):79–95 (2011).
12. Anastasopoulos, D., De Smedt, M., Vandewalle, L., De Roeck, G., and Reynders, E.P.B. "Damage identification using modal strains identified from operational fiber-optic Bragg grating data". *Structural Health Monitoring*, 17(6):1441–1459 (2017).
13. Jiang, X., Li, B., Mao, X., Peng, Y., and He, S. "New approach based on operational strain modal analysis to identify dynamical properties of the high-speed reciprocating operation mechanism". *Journal of Low Frequency Noise, Vibration and Active Control*, 38(3-4):1345–1362 (2019).
14. Wang, T., Celik, O., Catbas, F., and Zhang, L. "A frequency and spatial domain decomposition method for operational strain modal analysis and its application". *Engineering Structures*, 114:104–112 (2016).
15. Reynders, E., De Roeck, G., Gundes Bakir, P., and Sauvage, C. "Damage identification on the Tilff bridge by vibration monitoring using optical fiber strain sensors". *Journal of Engineering Mechanics*, 133(2):185–193 (2007).
16. Anastasopoulos, D., Moretti, P., Geernaert, T., De Pauw, B., Nawrot, U., De Roeck, G., Berghmans, F., and Reynders, E. "Identification of modal strains using sub-microstrain FBG data and a novel wavelength-shift detection algorithm". *Mechanical Systems and Signal Processing*, 86:58–74 (2017).
17. Wang, Z., Liu, M., Qu, Y., Wei, Q., Zhou, Z., Tan, Y., Hong, L., and Song, H. "The detection of the pipe crack utilizing the operational modal strain identified from fiber Bragg grating". *Sensors*, 19(11) (2019).
18. Anastasopoulos, D., Maes, K., De Roeck, G., Lombaert, G., and Reynders, E.P.B. "Influence of frost and local stiffness variations on the strain mode shapes of a steel arch bridge". *Engineering Structures*, 273:115097 (2022).
19. Murayama, H., Igawa, H., Omichi, K., and Machijima, Y. "Distributed sensing with OFDR and its application to structural health monitoring". *Proceedings of SPIE - The International Society for Optical Engineering*, 7753 (2011).
20. Liang, C., Bai, Q., Yan, M., Wang, Y., Zhang, H., and Jin, B. "A comprehensive study of optical frequency domain reflectometry". *IEEE Access*, 9:41647–41668 (2021).
21. Liehr, S., Nöther, N., and Krebber, K. "Incoherent optical frequency domain reflectometry and distributed strain detection in polymer optical fibers". *Measurement Science and Technology*, 21(1):017001 (2009).
22. Peña, F. "Fiber-optic strain-based deflection and twist sensing for a high-aspect-ratio swept wing". Technical Report TM-2020-220465, NASA (2020).
23. Richards, W.L., Parker Jr, A.R., Ko, W.L., Piazza, A., and Chan, P. "Application of fiber optic instrumentation". RTO-AG-160, Vol. 22, SCI-228, NATO Research and Technology Organization (RTO) (2012).
24. Dally, J.W. and Riley, W.F. *Experimental Stress Analysis*. McGraw-Hill, New York, 3rd edition (1991).
25. Kang, L.H., Kim, D.K., and Han, J.H. "Estimation of dynamic structural displacements using fiber Bragg grating strain sensors". *Journal of Sound and Vibration*, 305(3):534–542 (2007).
26. Peña, F., Park, B.C., and Chin, A. "Examination of doppler fiber optic sensing for structural velocity estimations". Technical Report TM-20210025579, NASA (2021).
27. Jutten, C. and Herault, J. "Blind separation of sources, part I: An adaptive algorithm based on neuromimetic architecture". *Signal processing*, 24(1):1–10 (1991).
28. Belouchrani, A., Abed-Meraim, K., Cardoso, J.F., and Moulines, E. "A blind source separation technique using second-order statistics". *IEEE Transactions on signal processing*, 45(2):434–444 (1997).
29. Siemens PLM. *FEMAP User Guide* (2020).
30. Allemang, R.J. "A correlation coefficient for modal vector analysis". In *Proceedings of the 1st International Modal Analysis Conference (IMAC)*, pages 110–116 (1982).

Chapter 2
Damping Ratio Assessment for Two Full-scale Operational Onshore Wind Turbines

T. Nunez and C. E. Ventura

Abstract With the increasing deployment of wind energy projects in highly seismic regions, new codes and recommended practice documents have emerged, such as NCh2369-2023 (Chile) and DNV-RP-0585-2021 (International), both influenced by IEC 61400-1 (2019) and IEC 61400-6 (2020). These documents consider the unique condition where seismic and wind loads are comparable. However, despite the new challenge imposed by this condition, added to a thriving industry that is constantly pushing to increase the size and height of wind turbines, most of the codes establish a value of 1% for a stopped turbine, that is, purely structural damping, and 5% for an operating turbine, a value that accounts for the aerodynamic behaviour associated with the rotation of the blades in a fluid. These values are based on limited literature, such as Prowell et al. [5] or Ishihara [6] for onshore WTG types. Nowadays, commercial turbines operating worldwide typically surpass 5 MW and 140 m hub heights. Hence, the need for continuous calibration of normative damping parameters, a key aspect of our ongoing research, appears evident. Based on the previously stated need, this work presents the experimental determination of the natural frequencies and their respective damping ratios for parked and operational conditions of two full-scales operational onshore WTG located in Chile and Argentina, with 140 mts and 87.6 mts, respectively. New correlations between period and damping ratio, including fore-aft and side-to-side directions, are proposed to update the current recommendation of the IEC61400-6 (Annex P) standard for design values.

Keywords Damping ratio · full-scale wind turbine · structural monitoring · onshore wind turbines

Introduction

Wind energy is one of the most rapidly growing renewable sources worldwide, having an installed capacity of 899 GW by 2022, triple that of a decade ago [1]. Much of this development has occurred in seismic zones, so the study of the structural and seismic design of wind turbines constitutes a key field not only for seismic engineering but also for global decarbonization goals. Although it is possible to use full FEA or multibody models, the most widely used model for preliminary seismic analysis of wind turbines corresponds to the "stick model," which simplifies the system, assuming it as a fixed base tower, with the mass of the equipment and blades lumped at its top and uniformly distributed through the tower, thus ignoring the displacement and flexibility of large components. This simplified model allows for calculating seismic loads using design spectra, readily available in most design codes. To perform this analysis, a proper determination of modal properties of the studied system, namely natural frequencies and damping, is essential to obtain accurate, representative results through the modal spectral method. In the case of damping for wind turbine seismic analysis, the most widely used standards, namely IEC61400 [2] and ASCE-AWEA RP2011 [3], establish a value of β=1% for a parked turbine, that is, purely structural damping, and β=5% for an operating turbine. This value accounts for the aerodynamic behaviour associated with the rotation of the blades in a fluid, in this case, air. These damping values are based on limited literature regarding full-scale tests available at the time of publication of the abovementioned standards, such as Prowell et al. [5], using a 65 KW, laboratory-tested, 21.9-meter hub height wind turbine or Ishihara [6] with a 36-meter hub-height full scales onshore type. Nowadays, commercial turbines operating worldwide typically go over 5 MW and 140 m hub heights. Hence, the need for continuous calibration of normative damping parameters appears evident. This study seeks to examine the results of ambient and operational vibration monitoring for two full-scaled onshore wind turbines located in southern Chile (140 mts hub-height), considered a

T. Nunez · C. E. Ventura
The University of British Columbia. Department of Civil Engineering
e-mail: nuneztom@mail.ubc.ca; ventura@civil.ubc.ca

Brandon Dilworth et al. (eds.), *Topics in Modal Analysis & Parameter Identification II, Vol. 10*, Conference Proceedings of the Society for Experimental Mechanics Series, https://doi.org/10.13052/9788743801559-2

seismic prone area, and Argentina (87.6 mts hub-height), evaluating their natural frequencies (periods) and damping ratios. Complementary, this study intends to validate the design damping ratio values recommended by IEC 61400-6 and propose an update of the current recommendation given in Annex P.

Literature Review

As previously stated, design codes [2, 3] coincide with the recommendations of damping values for the operational and seismic analysis of wind turbines. These codes establish damping ratio values associated with four operational conditions to be analyzed jointly with the earthquake case, as summarized in Table 1 below.

Table 1 Damping Values for seismic design under different operational conditions [2–4]

Operational Condition	Damping Ratio (β)
Standstill	1%
Normal operation, Idling and Emergency Stop	5%

In addition to the recommended design values previously summarized, IEC61400-6 [13], Annex P, presents a guideline for the adjustment of structural damping without aerodynamic effects -that is, only the standstill case- as a function of the first natural period (T1) of a steel tower as follows:

$$\beta\,(\%) = 2.0e^{-1.3\bullet T1} + 0.15 \tag{1}$$

It should be noted that the above expression is validated by different authors, such as Prowell [5], Ishihara [6] and Oh and Ishihara [7], but only calibrated for natural periods up to 2.8 seconds, using only 3 points. Figure 1 shows the plot and correlation derived from the Eq. (1).

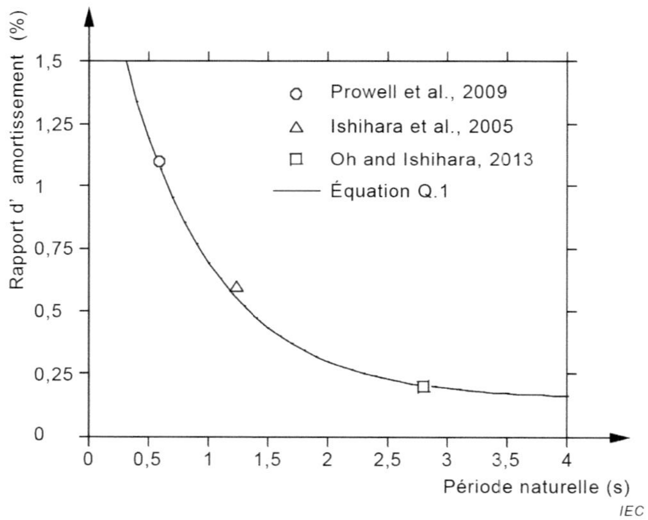

Fig. 1 First mode damping ratio for the steel tower of the wind turbine. IEC 61400-6

Oliveira et al. [11] long-term monitoring research reveals variability in damping ratios, correlating them with wind speed. They find an important increase in fore-aft damping and less impact on side-to-side damping.

Description of the Full-scale Test Case Studies

Case Study 01: 140-meter hub height, 4.5 MW, located in Chile

The test article consists of a commercial 4.5-MW wind turbine generator with a hub height of 140 m, already operational in Southern Chile, one of the most seismic regions in the world. It is worth mentioning that the site is in the rupture zone of two of the largest earthquakes in recent record (Valdivia 1960, Mw=9.5 and Maule 2010, Mw=8.8). Table 2 summarizes this turbine's main relevant design parameters, while Fig. 2 presents some images of the WTG and surroundings.

Table 2 Case Study 01. Wind Turbine Characteristics

Parameter	Value
Nominal power	4.5 MW
Hub height	140 m
Rotor diameter	150 m
Rotor and nacelle weight	210 t
Tower weight	390 t
Tower material	Steel
Tower configuration	5-sections

Fig. 2 Wind turbine generator, located in Chile. General dimensions

Case Study 02: 87.6-meter hub height, 4.5 MW, located in Argentina

The test article consists of a commercial 4.3-MW wind turbine generator with a hub height of 87.6 m, already operational in the province of San Luis, Argentina. Table 3 summarizes this turbine's main relevant design parameters, while Fig. 3 presents some images of the wind turbines and surroundings.

Table 3 Case Study 02. Wind Turbine Characteristics

Parameter	Value
Nominal power	4.5 MW
Hub height	87.6 m
Rotor diameter	150 m
Rotor and nacelle weight	210 t
Tower weight	207.1 t
Tower material	Steel
Tower configuration	3-sections

Fig. 3 Wind turbine generator, located in Argentina. General dimensions

Field Test Setup

Case Study 01: 140-meter hub-height

Experimental testing was performed on-site during a 2-session campaign, comprised of 2-hour standstill and 19-hour operational vibration measurements, including idling, stopping and continuously varying wind speeds, using four high-sensitivity velocity sensors and four acceleration sensors, as detailed in Table 4. Figure 4 shows the configuration for the monitoring plan. Only Trominos sensors were used for dynamic parameter estimation as a preliminary assessment.

Table 4 Vibration Measurement Equipment

Equipment	Quantity	Characteristics
High-sensitivity velocity sensors	4 (one at the tower bottom, one at mid-height and two at the tower top). Triaxial.	Moho Tromino @128 Hz
Acceleration sensors	4 (one at tower bottom, two at middle positions and one at tower top). Triaxial	Endaq S3 @128 Hz

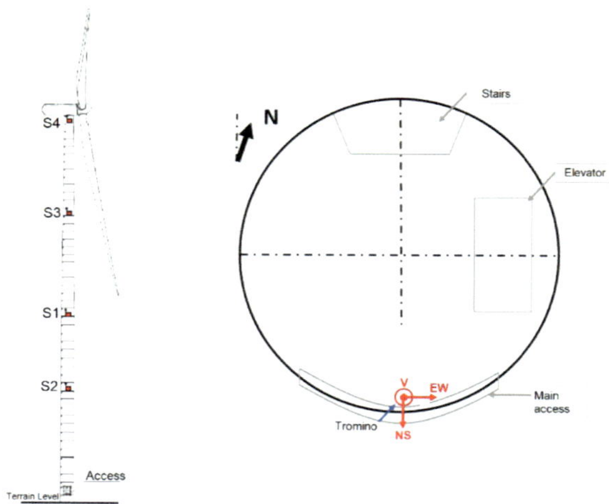

Fig. 4 Monitoring plan (Tromino only). Left: Sensor deployment along the tower. Right: sensor location in plan (typical)

Case Study 02: 87.6-meter hub-height

Experimental testing was performed on-site during a 2-session campaign, comprised of 18-hour operational vibration measurements, including idling, standstill, and continuously varying wind speeds, using four high-sensitivity velocity sensors, as detailed in Table 5. Figure 5 shows the configuration for the monitoring plan.

Table 5 Vibration Measurement Equipment

Equipment	Quantity	Characteristics
Acceleration sensors	3 (two at the tower top and one at the middle position). Triaxial	Endaq S3 @256 Hz

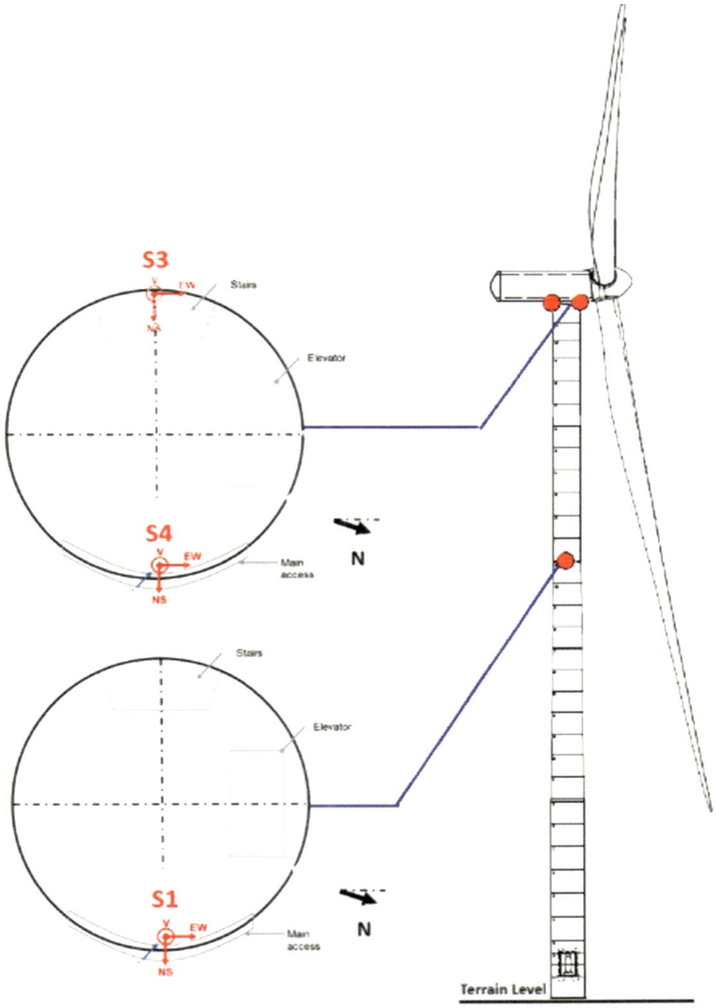

Fig. 5 Monitoring plan (Tromino only). Left: Sensor deployment along the tower. Right: sensor location in plan (typical)

Dynamic Parameters Results

The frequency domain decomposition method and enhanced frequency domain decomposition method were adopted to extract modal frequencies from ambient vibration data. For damping ratio identification, two methods were considered: The half-power method and the Half-quadratic gain method [9].

Parked Condition Results

Table 6 Identified Modal Parameters – Standstill Case. Case Study 01: 140-meter hub height

N°	Frequency [Hz]	Period [s]	β [%]	Direction
1	0.14	6.98	1.25	FORE-AFT
2	0.14	6.98	1.38	SIDE-SIDE
3	0.56	1.79	1.84	SIDE-SIDE
4	0.67	1.49	0.68	FORE-AFT
5	1.08	0.93	0.94	SIDE-SIDE
6	1.16	0.86	0.18	FORE-AFT
7	1.54	0.65	0.39	FORE-AFT

Table 7 Identified Modal Parameters – Standstill Case. Case Study 02: 87.6-meter hub-height

N°	Frequency [Hz]	Period [s]	β [%]	Direction
1	0.23	4.44	0.80	FORE-AFT
2	0.23	4.44	0.94	SIDE-SIDE
3	0.82	1.22	0.35	COUPLED
4	0.88	1.14	1.02	COUPLED
5	2.16	0.46	0.39	COUPLED
6	2.25	0.44	1.11	COUPLED
7	3.16	0.32	0.79	SIDE-SIDE

Operational Condition Results

Operational conditions is only reported for the 140-meter-high WTG located in Chile. Once the initial standstill measurement was completed, a 19-hour measurement was performed with the turbine under normal operational conditions, which included various idling, stopping, and varying wind speed/rotor angular velocity conditions. Operational data, such as average ambient windspeed and rotor angular velocity, were obtained from the turbine's SCADA[1] system, Fig. 6 provides relevant information to correlate with the modal parameters, mainly associated with variations in the wind turbines' damping ratio.

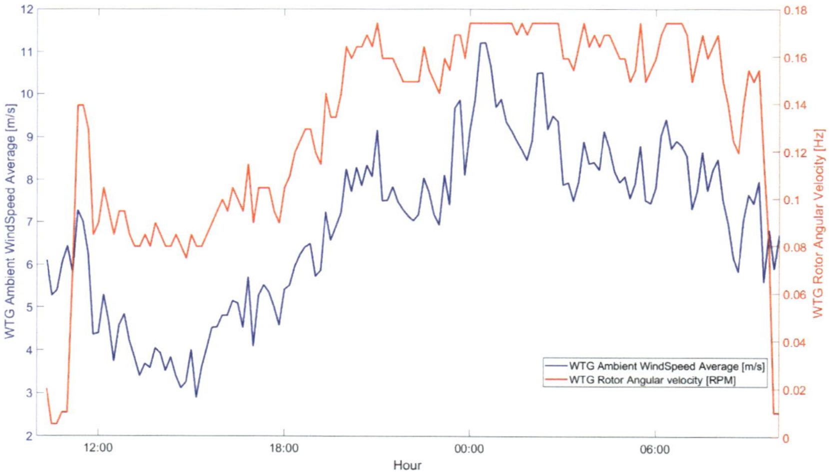

Fig. 6 WTG ambient wind speed average and rotor angular velocity. Case Study 01: 140 mts hub-height

[1] Supervisory Control and Data Acquisition

Considering the duration of the monitoring and the combination of natural and operational frequencies given by the rotor-blades system, the use of the variation of frequency over time constitutes a useful tool to visualize variations in frequency values, as well as closer frequencies during transient periods. In this case, a spectrogram is presented in Fig. 7 to evaluate frequency behaviour and interaction between frequencies over time.

From Figs. 6 and 7, the rotor angular frequency variation reported by the SCADA system coincides with the spectrogram in terms of the variation form. However, the spectrogram presents all the variations amplified by a factor of 3, typically called 3P, associated with three-per-revolution frequency, considering the WTG has three blades. Additionally, the operational 1P frequency seems to be close to the first and second natural frequency of the WTG (fore-aft and side-to-side, first bending). However, most wind turbines work in a range of 1P (or 3P) value, with complex control systems to monitor and bypass resonance effects, e.g., by adding a notch filter to prevent the rotor from resonating at a structural natural frequency for an extended period.

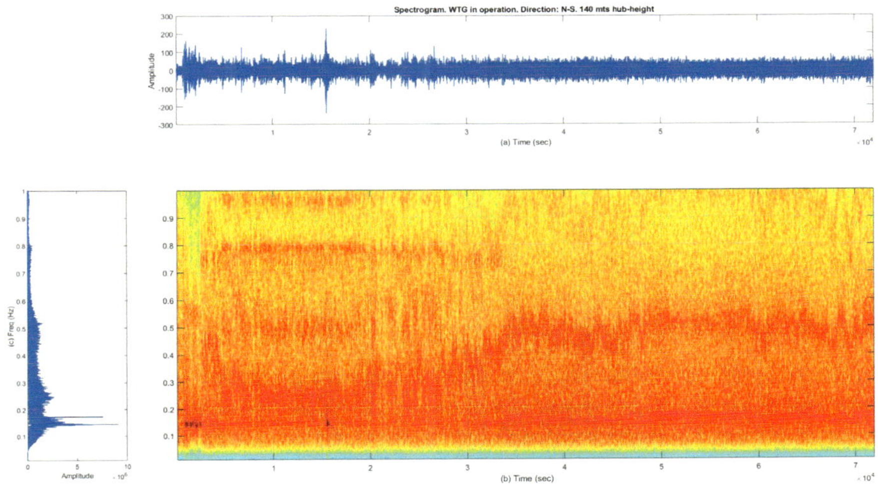

Fig. 7 Spectrogram for 19-hour operational conditions. E-W component (fore-aft.). Case Study 140 meters hub-height. Chile

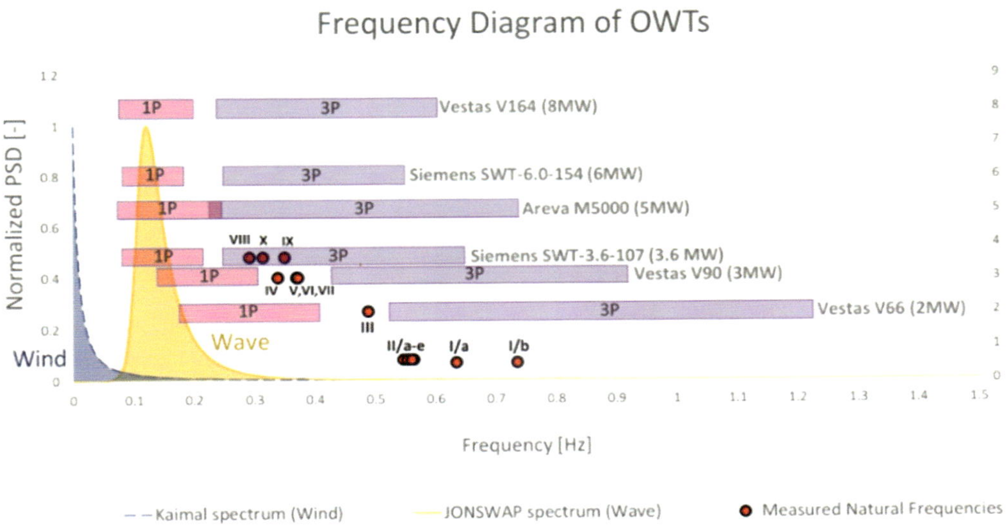

Fig. 8 Typical wind and wave spectra, rotational speed (1P) and blade passing (3P) frequency bands for commercial turbines and measured natural frequencies of the WTG [14]

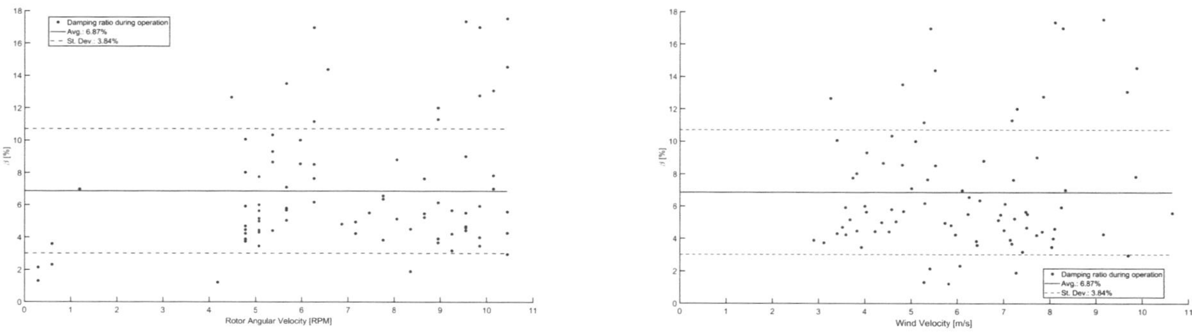

Fig. 9 (a) Damping ratio for operational condition for different (left) rotor angular velocity and (rigth) wind velocity. Only the first bending fore-aft mode is presented

Given the proximity between the operational and natural frequencies of the system, assessing operational damping using the methodologies outlined in this study proves to be complex. Nonetheless, judiciously selecting bandpass filters makes it feasible to isolate the frequency and compute damping across various wind speeds and/or turbine rotor angular frequencies. Figure 9 represents the scattered damping ratio for different wind speeds and rotor angular velocities, aiming to discern any correlation or trend. Each point was obtained for a 30-minute time series processed in the frequency domain and filtered close to the first bending fore-aft mode to obtain the operational damping ratio using the methodologies presented in this article.

From the previous figure, it is unclear whether a trend correlating the increase or decrease in damping with the increase in wind speed or rotor angular velocity can be discerned. However, the results confirm that using a 5% value for operational damping is reasonable. From the figure, an average damping value of 6.87% is obtained, with a standard deviation of σ=3.84%. It should be noted that to obtain a better estimation, some of the following actions are required: (a) longer monitoring periods to increase the sample size, with the additional benefit of generating a structural health monitoring system, and (b) more complex and effective methodologies for extracting operational damping ratios by separating natural frequencies, such as the Hilbert-Huang Transform (HHT) or the wavelet method.

Conclusion

From the analyses of the ambient vibration tests, it was possible to determine frequencies and damping ratios values indistinctly using velocity (Trominos) and acceleration (Endaq) measurements. This is mainly due to the wind conditions that excite the WTG over the noise level of the instruments, especially for acceleration measurements (less sensitive than the velocity measurements).

Using the damping ratios obtained from the standstill measurements and the literature review, it is possible to validate a damping design parameter for parked conditions. Although, it is recognized that there is a certain dependency of the period (and therefore, the mass and stiffness) of the system.

This study shows that currently recommended damping values, such as recommended in Annex P of IEC 61400-6 must be reviewed. Furthermore, the equation presented in that standard does not define the directionality of the natural period, typically known as fore-aft and side-to-side.

In addition, while the value indicated by Prowell [10] is accurate, it corresponds to a scale model developed for laboratory testing. The Ishihara [6] value used for the damping ratio presented in Fig. 1 is for side-to-side mode; however, the value for the fore-aft damping ratio is 1%. Lastly, the value of the damping ratio presented by Oh and Ishihara [7] was obtained from an offshore wind turbine with a tuned-mass damper inside the tower. This particular case cannot be directly compared with an onshore configuration.

It should be noted also that the equation proposed in Annex P of IEC 61400-6, tends to $\beta = 0.15\%$ for longer periods (over 4 seconds), and the value obtained from this study differs significantly from the results obtained from full-scaled WTG monitoring.

It is, therefore, reasonable to propose an updated formula for the damping estimation using the results from this and previous studies, such as those by Ishihara [6] and Oliveira [11]. The following Tables present the damping rations for different periods, for on-shore, steel-tower and distinguishing between fore-aft and side-to-side damping ratio:

Table 8 First Mode fore-aft direction damping ratio (β) comparison from different authors

N°	Author / (Publication Year)	Hub-Height [m]	Rotor Diameter [m]	Natural Frequency [Hz]	Period [s]	β [%]
1	Ishihara (2005)	36.0	31.0	0.81	1.23	1.00%
2	Oliveira et. al. (2018)	76.0	82.0	0.35	2.84	0.68%
3	This Study (2024)	87.6	150.0	0.22	4.55	0.80%
4	This Study (2024)	140.0	150.0	0.14	7.14	1.25%
				Average		0.93%
				Std. Dev.		0.25%

Table 9 First Mode side-side direction damping ratio (β) comparison from different authors

N°	Author / (Publication Year)	Hub-Height [m]	Rotor Diameter [m]	Natural Frequency [Hz]	Period [s]	β [%]
1	Ishihara (2005)	36	31	0.81	1.23	0.60%
2	Oliveira et. al. (2018)	76	82	0.35	2.83	0.80%
3	This Study (2024)	87.6	150	0.22	4.55	0.94%
4	This Study (2024)	140	150	0.14	7.14	1.38%
				Average		0.93%
				Std. Dev.		0.33%

As observed from Tables 8 and 9, it seems reasonable to establish a correlation between the natural period and the damping ratio, updating the equation presented in Annex P of IEC 41600-6. Graphically, Figs. 8 and 9 display the table results, with the average value and its corresponding standard deviation.

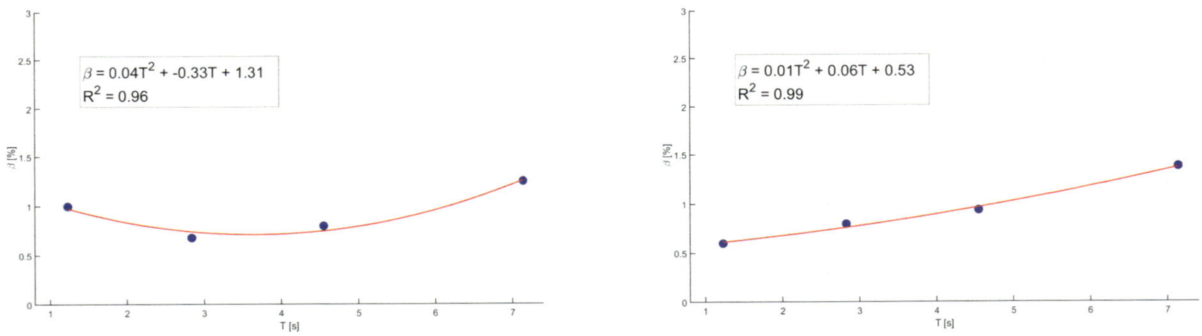

Fig. 10 Proposed relation between damping ratio as a function of the first bending fore-aft natural period (left) and side-to-side natural period (right)

These correlations are subject to change as more towers are assessed, reflecting the evolving landscape of the wind energy market characterized by taller towers and increased power outputs.

Regarding operational vibration measurements, it was not clear to correlate directly the results obtained for the damping ratio with the rotor's angular velocity or wind velocity provided by WTG's SCADA. Despite this fact, the values obtained have an acceptable mean value of 6.78%, close to the design value of 5% proposed by most of the design standards.

The study revealed the difficulties associated with identifying modal parameters when operational frequencies are inherent to the equipment that are also close to the natural frequencies. This is mainly due to the coupling of frequencies, which does not allow for precise damping estimation using the methods employed in this work (half power). From this perspective, it is necessary to explore more complex techniques, such as the Hilbert-Huang Transform (HHT) and wavelet analysis. Additionally, this type of study paves the way for developing continuous monitoring systems, whether operational or seismic.

Acknowledgments We extend our acknowledgment to Ricardo Mendez and Angel Vidal for their support and assistance in conducting the full-scaled field tests. Their expertise and dedication significantly contributed to the success of this research.

References

1. IRENA (2023), Renewable energy statistics 2023, International Renewable Energy Agency, Abu Dhabi.
2. International Electrotechnical Commission (2019). IEC 61400-1 Edition 4 Wind Energy Generation Systems. Part 1: Design Requirements.
3. ASCE-AWEA (2011).RP2011 Recommended Practice for Compliance of Large Land-based Wind Turbine Support Structures, American Society of Civil Engineers-
4. INN (2003). NCh.2369 Of.2003. Diseño Sísmico de Estructuras e Instalaciones Industriales, Instituto Nacional de Normalización.
5. Prowell, I., Veletsos, M., and Elgamal, A., (2008). "Full Scale Testing for Investigation of Wind Turbine Seismic Response." Proceedings of the 7th World Wind Energy Conference, Kingston, Ontario, Canada.
6. Ishihara, T., Phuc, P. V., Fujino,Y., Takahara, K., Mekaru, T.(2005), A Field Test and Full Dynamic Simulation on a Stall Regulated Wind Turbine, Proceedings of APCWE VI, p.599-612.
7. Oh, S., Ishihara, T. (2018), Structural parameter identification of a 2.4MW bottom Fixed wind turbine by excitation test using an active mass damper, Wind Energy, 21(11), 1232-1238.
8. Knudsen, T., Bak, T., & Tabatabaeipour, S. (2012). Detection of excessive wind turbine tower oscillations fore-aft and sideways. 2012 American Control Conference (ACC), 5795-5800.
9. Casiano, M.J. (2016). Extracting Damping Ratio from Dynamic Data and Numerical Solutions.
10. Prowell, I. (2011). An experimental and numerical study of wind turbine seismic behavior. UC San Diego. ProQuest ID: Prowell_ucsd_0033D_11508. Merritt ID: ark:/20775/bb9660041c. Retrieved from https://escholarship.org/uc/item/82b829mg
11. Gustavo Oliveira, Filipe Magalhães, Álvaro Cunha, Elsa Caetano (2018). Continuous dynamic monitoring of an onshore wind turbine, Engineering Structures, Volume 164, Pages 22-39, ISSN 0141-0296, https://doi.org/10.1016/j.engstruct.2018.02.030.
12. IEC 61400-6 Edition 1 2020-04. Wind Energy Generation Systems. Part 6: Tower and Foundation Design Requirements., International Electrotechnical Commission., 2020.
13. DNVGL-RP-0585. Seismic Design of Wind Power Plants. DNV-GL Recommended Practice, DNV-GL, 2021.
14. Arany, Laszlo & Bhattacharya, Subhamoy & Macdonald, John & Hogan, Stephen. (2016). Closed-form solution of Eigen frequency of monopile supported offshore wind turbines in deeper waters incorporating stiffness of substructure and SSI. Soil Dynamics and Earthquake Engineering. 83. 10.1016/j.soildyn.2015.12.011.

Chapter 3
Effects of Environmental and Operational Variables on the Modal Properties of Wind Turbines Blades

Felipe Bravo and Rodrigo Astroza

Abstract This research addresses the influence of environmental and operational variables on modal properties of wind turbine blades by employing a vibration-based structural health monitoring approach. Three triaxial sensors were placed in a mock-up resin/epoxy wind turbine blade and exposed to different rotor speeds, mass additions, and progressive damage under actual environmental conditions. Modal properties were identified with an input-output state-space system identification algorithm and an automated modal tracking approach. The findings reveal that higher rotor speeds increase the natural frequency of the first flapwise mode, attributed to greater blade stiffness. In contrast, higher temperatures significantly decrease the natural frequency, particularly at higher speeds, while added mass lowers natural frequency, with diminishing effects at increased speeds. Notably, although temperature has a pronounced impact on frequency in early damage stages, progressive damage did not significantly influence damping ratios or mode shapes. These insights underscore the intricate relationship between environmental and operational variables and blade performance, highlighting the necessity for improved monitoring strategies in operational modal analysis.

Keywords Structural Health Monitoring · Wind Turbine Blades · Environmental and Operational Variability · Modal Analysis

Introduction

Wind energy is increasingly recognized as a key player in the transition towards sustainable energy systems, providing a clean and renewable source of energy and being one of the fastest-growing forms of renewable energy. Despite the advancements in wind energy technology, wind turbine blade failures remain a prevalent issue [1], leading to significant economic losses, reduced operational efficiency, and serious safety concerns. Blade failures can result from a variety of factors, including material fatigue, manufacturing defects, and environmental stressors [2]. The economic implications of such failures are substantial, often involving costly repairs, downtime, and potential legal liabilities. Moreover, failures can compromise safety, posing risks to maintenance personnel and nearby communities. Addressing these challenges is imperative to ensure the continued growth and viability of wind energy as a primary energy source.

While there is a growing body of literature on wind turbine blade performance, there remains a significant gap in comprehensive studies examining how environmental and operational variables affect blade performance and durability. Most existing research has focused on isolated factors or laboratory conditions [3], failing to account for the complex interplay of multiple variables encountered in real-world operating environments. This lack of holistic understanding hampers the development of effective monitoring and maintenance strategies, which are crucial for enhancing blade longevity and reliability.

This research investigates the influence of key environmental and operational variables—specifically temperature, rotor speed, and mass changes—on the modal parameters of wind turbine blades. By employing a vibration-based methodology, this study aims to identify how these variables interact and affect the structural integrity of the blades. Ultimately, the findings of this research will contribute to a deeper understanding of the factors influencing blade performance, facilitating the development of improved monitoring and maintenance practices to enhance the reliability and longevity of wind turbine systems.

Felipe Bravo · Rodrigo Astroza
Faculty of Engineering and Applied Sciences, Universidad de los Andes, Santiago, Chile
e-mail: fbravo1@miuandes.cl; rastroza@miuandes.cl

Brandon Dilworth et al. (eds.), *Topics in Modal Analysis & Parameter Identification II, Vol. 10*, Conference Proceedings of the Society for Experimental Mechanics Series, https://doi.org/10.13052/9788743801559-3

Background

Vibration-based structural health monitoring (SHM) approaches employing modal parameters—such as natural frequencies, damping ratios, and mode shapes—are essential for assessing environmental and operational variables (EOVs) and damage-sensitive features used in damage detection [4, 5]. SHM leverages these modal parameters as critical indicators of structural integrity, particularly in wind turbine blades, which are subjected to dynamic loads and varying environmental conditions. Research has demonstrated that variables such as temperature [1], rotor speed [2], and ambient noise [6] significantly affect the modal parameters of wind turbine structures, particularly of blades. Both numerical and experimental studies have revealed how damage impacts these modal parameters [7], complicating damage detection through vibration analysis alone. This complexity can lead to misdiagnosis, resulting in unnecessary maintenance or, conversely, unaddressed structural issues that could compromise safety.

Operational modal analysis (OMA) must navigate the complexities arising from the interaction of EOVs, as the influence of damage on modal parameters can be confounded by these environmental factors. For instance, high ambient noise levels can mask significant changes in modal parameters, making it difficult to detect damage reliably. Mitigating the impact of EOVs on modal parameters is critical for effective OMA, especially since damage diagnosis and prognosis models—such as the one developed by Jaramillo et al. [8]—often rely on natural frequencies as key inputs. Natural frequencies are particularly significant, as they serve as the primary indicators for assessing structural health and predicting potential failures.

Data acquisition on wind turbines presents its own challenges. While the tower structure is generally easier to sensorise [9], obtaining reliable data from the blades remains problematic. Factors such as blade shape, material properties, and environmental exposure contribute to the difficulty in effective sensor deployment. Most existing data has been collected under laboratory conditions, simulations [6, 10], or through methodologies that are too invasive to replicate [11]. These limitations raise concerns about the applicability of the findings in real-world settings, where environmental conditions vary widely.

This study proposes a vibration-based methodology for accurately identifying modal parameters using input-output system identification and effectively clustering these parameters under varying EOVs. By integrating these factors, this research aims to improve the reliability of damage detection models and inform maintenance strategies. It presents a data framework that accounts for variations in the modal parameters of wind turbine blades due to changes in temperature and rotor speed and describes the combined effects of these variables on the first and second modes of the blades. Additionally, this study examines the influence of progressive damage on the modal parameters of a small-scale wind turbine blade. The findings could provide significant insights for the wind energy sector, contributing to enhanced monitoring practices and longer turbine lifespans.

Analysis

Acceleration measurements were captured from a 2.4 m long glass-fiber/epoxy-resin wind turbine blade from a wind turbine mock-up designed specifically for this study. A 3-meter-tall steel tube with a 1 hp motor attached to the top, controlled by a frequency regulator, and equipped with a hexagonal steel rotor allowed for better control of the operational variables. This setup facilitated precise adjustments, as illustrated in Fig. 1.

Fig. 1 Experimental Wind Turbine Setup

Three strategically deployed ADXL345 accelerometers, featuring an amplitude range of ±8 g and a sensitivity of 2 mg/LSB, were installed on the blade, as shown in Fig. 2. A monitoring device located in the center of the wind turbine's rotor captured temperature and acceleration data at a sampling rate of 200 Hz. Measurements were taken every 10 minutes under various scenarios, including healthy states, changes due to mass additions, and progressively induced artificial damage while exposing the blade to varying rotor speeds. These scenarios were analyzed to assess the impact of EOVs on the wind turbine blade, represented in Fig. 3.

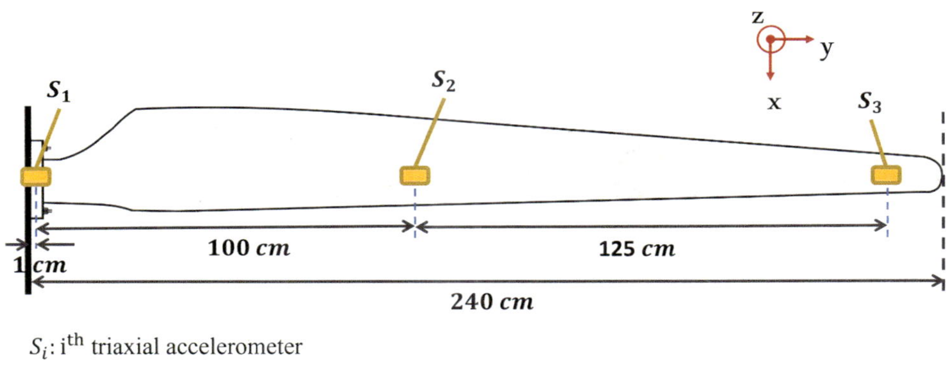

Fig. 2 Accelerometer deployment on the blade

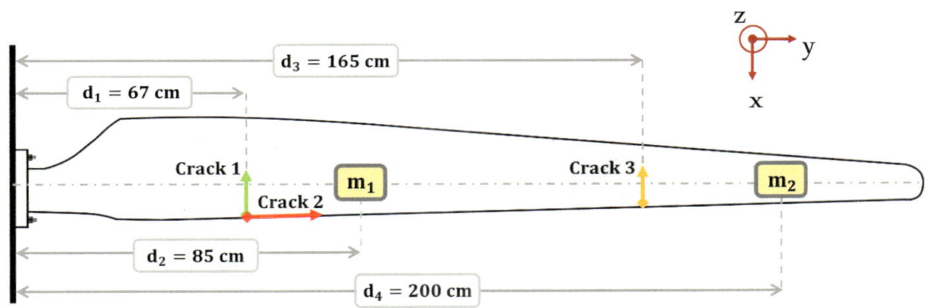

Fig. 3 Scenarios for testing EOVs

An input-output system identification method, specifically the Deterministic-Stochastic Identification (DSI) algorithm [10], was employed to estimate the modal parameters, including natural frequency, damping ratio, and mode shape of the wind turbine blade. An automated modal tracking algorithm then enabled accurate tracking of the blade's modal properties throughout the different tested scenarios over several days of data collection, capturing a diverse range of temperature fluctuations and ambient conditions.

A significant ascending trend in the natural frequency of the first flapwise mode was observed as the rotor speed increased (Fig. 4). This effect is attributed to heightened blade deformation and increased stiffness at higher speeds due to nonlinear geometry effects. The damping ratio exhibited a slight increase at lower speeds ($\leq$ 45 rpm) but showed a substantial rise at 60 rpm. Despite these changes in frequency and damping, no significant variation in mode shape was detected across different rotor speeds.

Higher temperatures resulted in a decrease in the natural frequency of the first flapwise mode, with steeper slopes observed at increasing rotor speeds (Fig. 5). The damping ratios exhibited a stronger temperature influence at speeds above 15 rpm. At lower speeds, other factors appeared to outweigh the temperature effects on damping behavior (Fig. 5). The addition of mass significantly lowered the natural frequency of the blade, with this effect becoming less pronounced at higher rotor speeds due to compensation effects (Fig. 6).

In the earlier stages of damage, temperature had a more pronounced effect on the natural frequency, allowing for noticeable damage influences. However, in later stages of damage, a significant decline in frequency was observed as damage progressed as rigidity tends to decrease over damage progresses. Despite visible frequency effects, no clear influence of damage on the damping ratio or mode shape was detected (Fig. 7).

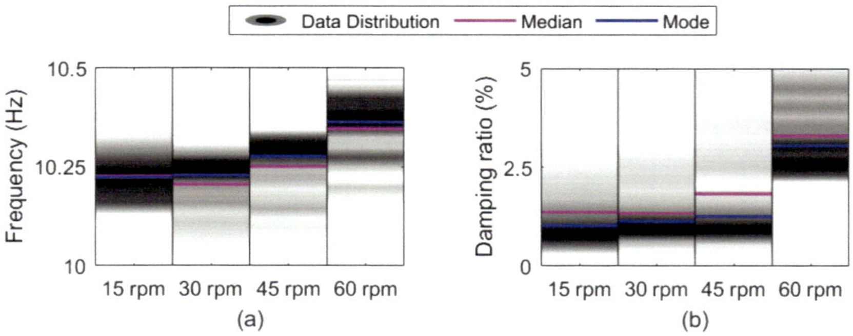

Fig. 4 (a) Identified frequency and (b) damping ratio at rotating speed scenarios

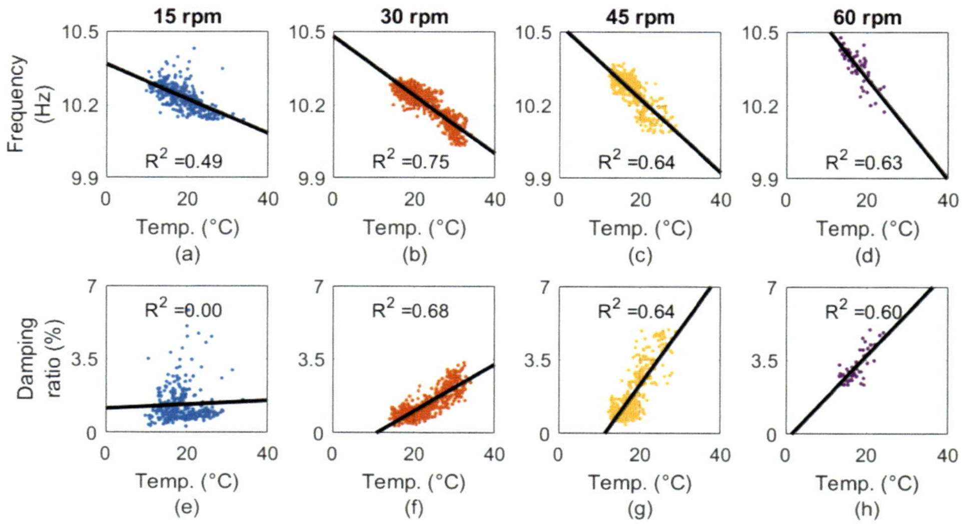

Fig. 5 Identified frequency at four rotating speed scenarios (a), (b), (c) and (d) and identified damping ratio at four rotating speed scenarios (e), (f), (g) and (h)

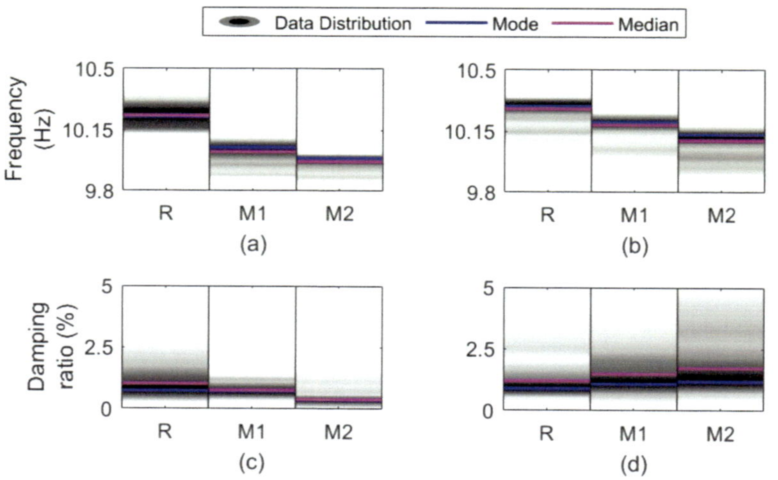

Fig. 6 Identified frequency considering reference and mass addition scenarios at 15 rpm (a) and 45 rpm (b) and identified damping ratio considering reference and mass addition scenarios at 15 rpm (c) and 45 rpm (d)

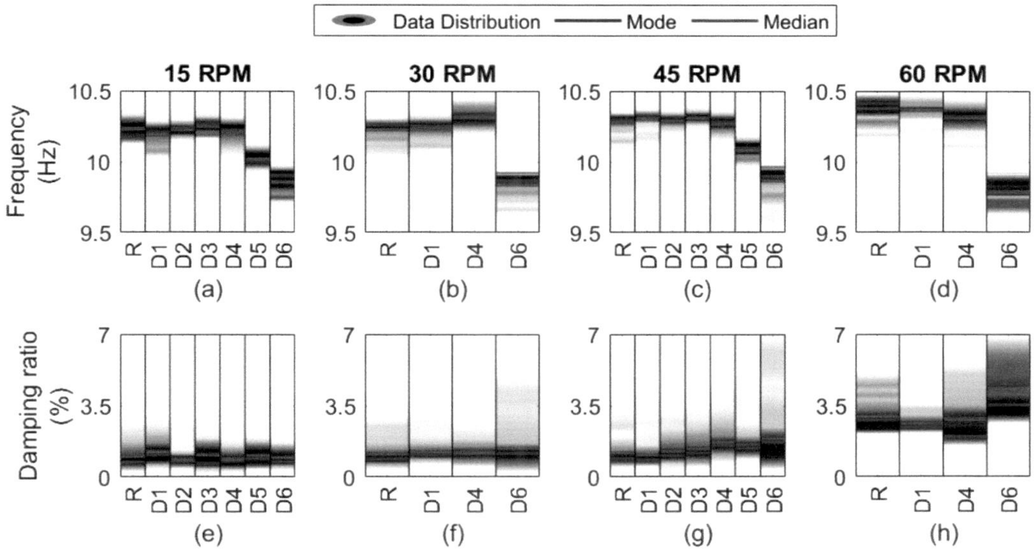

Fig. 7 Identified frequency at four rotating speed scenarios (a), (b), (c) and (d) and identified damping ratio at four rotating speed scenarios (e), (f), (g) and (h) considering progressive damage

The results align with the established physics related to material behavior and corroborate findings from previous research on damage progression in rotor blades [6]. These insights into the relationship between modal properties and EOVs are crucial for enhancing OMA. Algorithms based on modal properties could benefit from these findings to better understand, compensate, and disaggregate the effects of each environmental and operational variable, allowing for improved isolation of the real effects on wind turbine performance.

Conclusion

This study has established significant correlations between environmental and operational variables and the modal parameters of wind turbine blades. Key findings include that the natural frequency of the first flapwise mode increases with rotor speed, reflecting enhanced stiffness due to nonlinear geometric effects. Conversely, elevated temperatures have been shown to decrease natural frequency, especially at higher rotor speeds. The addition of mass to the blades significantly reduces natural frequency, although this effect becomes less pronounced at higher speeds. Additionally, while temperature influences the modal parameters during the early stages of damage, no substantial impact on damping ratios or mode shapes was observed as damage progressed. These findings highlight the complex interplay between environmental and operational variables and the structural dynamics of wind turbine blades.

Understanding how environmental and operational variables affect modal properties, maintenance strategies can be optimized, leading to more effective monitoring and timely interventions. Improved damage detection methods that incorporate these variables can help prevent catastrophic failures, thereby ensuring operational efficiency and reducing maintenance costs. This research contributes to the development of more robust operational modal analysis frameworks that account for the variability introduced by EOVs, ultimately supporting the sustainable operation of wind energy systems.

Acknowledgments The authors acknowledge the funding from the Chilean National Agency for Research and Development (ANID) through the projects Tecnologías Avanzadas TA23I10055 and FONDECYT Regular 1240503.

References

1. M. D. Reder, E. Gonzalez and J. J. Melero, "Wind Turbine Failures - Tackling current Problems in Failure Data Analysis," *Journal of Physics: Conference Series*, 2016.

2. W. Wang, Y. Xue, C. He and Y. Zhao, "Review of the Typical Damage and Damage-Detection Methods of Large Wind Turbine Blades," *Energies*, 2022.
3. Y. Ou, K. E. Tatsis, V. K. Dertimanis, M. D. Spiridonakos and E. N. Chatzi, "Vibration-based monitoring of a small-scale wind turbine blade under varying climate conditions. Part I: An experimental benchmark," *Structural Control and Health Monitoring*, 2020.
4. C. Aaskov Drangsfeldt and L. D. Avendaño, "A combined implicit-explicit vibration-based SHM method for damage detection of wind turbine blades," in *EuroDyn2023*, Delft, 2023.
5. S. Pereira, J. Pacheco, F. Pimienta, C. Moutinho, A. Cunha and F. Magalhaes, "Contributions for enhanced tracking of (onshore) wind turbines modal parameters," *Engineering Structures*, 2023.
6. J. Pacheco-Chérrez, D. Cárdenas, A. Delgado and O. Probst, "Operational modal analysis for damage detection in a rotating wind turbine blade in the presence of measurement noise," *Composite Structures*, 2024.
7. S. Kulkarni and S. Patil, "Numerical and experimental investigations in a cracked wind turbine blade using natural frequencies," *Wind Energy*, 2024.
8. F. Jaramillo, J. M. Gutierrez, M. Orchard, M. Guarini, and Astroza, R., "A Bayesian approach for fatigue damage diagnosis and prognosis of wind turbine blades," *Mechanical Systems and Signal Processing*, 2022.
9. M. Khazaee, P. Derian and A. Mouraud, "A comprehensive study on Structural Health Monitoring (SHM) of wind turbine blades by instrumenting tower using machine learning methods," *Renewable Energy*, 2022.
10. D. Tcherniak and G. C. Larsen, "Application of OMA to an Operating Wind Turbine: now including Vibration Data from the Blades," in *5th International Operational Modal Analysis Conference (IOMAC'13), Guimarães*, 2013.
11. A. Movsessian, D. García Cava and D. Tcherniak, "An artificial neural network methodology for damage detection: Demonstration on an operating wind turbine blade," *Mechanical Systems and Signal Processing*, 2021.

Chapter 4
Structural Dynamics Identification using Physics-Informed SINDy

Celso T. do Cabo, Raymond Joshua, and Zhu Mao

Abstract Inverse problems in engineering, such as performing system identification and obtaining the Partial Differential Equations (PDEs) of a system is challenging. While it is possible to correlate experimental measurements with a Finite Element Model (FEM) and use the adjusted FEM as a digital twin, it is difficult to obtain such parameters only from measurements. For instance, it is possible to obtain the natural frequencies, mode shapes and damping ratio of a vibrational system using traditional signal processing techniques, but the governing equations may not be easily obtained. With the increase application of machine learning techniques, a specific algorithm developed for system identification is the Sparse Identification of Nonlinear Dynamics (SINDy), which uses the measured signal and its derivatives to estimate the parameters of its PDEs in a matrix form. This work applies SINDy to identify the PDEs of a vibrational system, which consists of second order PDEs. Considering the nature of the problem, the technique is also updated to include physical knowledge of the structure, such as its total mass and first natural frequency, leading to the final estimation of the mass matrix to not be a normalized estimation, allowing the correct estimation of mass, stiffness and damping matrices, as well as a good reconstruction of the original measurement and correct estimation of mode shapes and natural frequencies from the obtained parameters.

Keywords Machine Learning · System Identification · Sparse Identification · Nonlinear Dynamics · Structural Dynamics

Introduction

Inverse problems in engineering, including system identification and obtaining the partial differential equations (PDEs) of a structure can be challenging. Many studies were performed aiming to identify system behavior for linear and nonlinear systems. Some of the approaches for dynamical systems can include regression models, neural networks and fuzzy logic [1, 2]. More recently, with the increase of neural network applications, the Physics Informed Neural Networks (PINNs), were also employed in system identification [3, 4]. However, among the linear regression types, Sparse Identification of Nonlinear Dynamics (SINDy) proposed by Brunton et al. shows great potential for system identification [5]. This technique requires measurements and its derivatives or integrals to obtain the governing equations of a system. Some of the applications for the technique include systems with one degree of freedom (DOF) [6, 7], or in rotary machinery for load identification [8]. Nonlinear normal modes (NNM) of discrete systems were also investigated using SINDy [9]. In addition, other studies were performed on systems with implicit dynamics [10] or in different methodologies to perform the optimization for SINDy, such as employing Runge-Kutta [11] or Gaussian process (GP) for force estimation [12].

However, vibrational systems pose higher complexity, since they possess second order PDEs behavior. Although SINDy has been employed for PDEs applications previously [13]. When dealing with structural dynamics, where continuous systems

NSC-614-5662 10/2024
Unclassified Unlimited Release

Celso T. do Cabo · Zhu Mao (✉)
Department of Mechanical and Materials Engineering Worcester Polytechnic Institute, Worcester, MA 01609
e-mail: ctdocabo@wpi.edu; zmao2@wpi.edu

Raymond Joshua
Honeywell Federal Manufacturing & Technologies, LLC operates the Kansas City, National Security Campus for the United States Department of Energy/National Nuclear, Security Administration under Contract Number DE-NA0002839
e-mail: rjoshua@kcnsc.doe.gov

Brandon Dilworth et al. (eds.), *Topics in Modal Analysis & Parameter Identification II, Vol. 10*, Conference Proceedings of the Society for Experimental Mechanics Series, https://doi.org/10.13052/9788743801559-4

will lead to dependent mass and stiffness, the estimation of the governing equations poses to be challenging. This work proposes a modification of the traditional SINDy, aiming to perform a system identification of a continuous vibrational system, based on sparse measurements. For that, the modification of the algorithm is given by modifying the linear regression to a stochastic gradient descent (SGD) and including penalty terms to the loss function such that the parameters obtained are encouraged to behave similarly to what would be observed from a finite element (FE) model after reducing its observations to the number of sensors obtained.

Theory Background

To obtain the partial differential equations (PDEs) of a system, the sparse identification of nonlinear dynamics (SINDy) originally proposes a linear regression from the measurements obtained and a library of possible combinations of parameters to obtain the governing equation. For the case of a second order system, the acceleration measurements $\left[\ddot{X}\right]$ is the result that the equation tries to obtain; by integrating the measurements it is possible to build the library of parameters $\Theta\left(\dot{X}, X\right)$ which will consist of the velocity and displacement measurements. By performing the linear regression as shown in Eq. (1), it is possible to estimate the parameters Ξ.

$$\Xi = \left(\Theta^T\Theta\right)^{-1}\Theta^T\ddot{X} \tag{1}$$

The parameters Ξ will then possess the stiffness and damping coefficients normalized by the mass of the system. However, this approach assumes that the mass is independent between the degrees of freedom (DOFs) of the structure and the mass matrix will be composed by an identity matrix. For a more realistic representation of the system, it is proposed a modification in the original algorithm so that the mass matrix is similar to what is expected for vibrational systems, the overall process to obtain the governing equations is presented on the flowchart in Fig. 1. To include the mass matrix in the library of parameters, it is required to modify the original SINDy model from a linear regression to an optimization form. For that, it was chosen to perform a stochastic gradient descent (SGD), and the new library of possible parameters will be changed to $\Theta\left(\ddot{X}, \dot{X}, X\right)$, so that it also has the mass parameters of the system. To optimize the model using SGD, the gradient of the mean squared error must be computed, and it is calculated based on Eq. (2).

$$\nabla_\xi f_{MSE}\left(\ddot{x}, \hat{\ddot{x}}, \xi\right) = \frac{1}{n}\Theta\left(\Theta^T\Xi + b - \ddot{X}\right) \tag{2}$$

where b is the bias of the model. Finally, the values of the parameters will be updated based on Eq. (3).

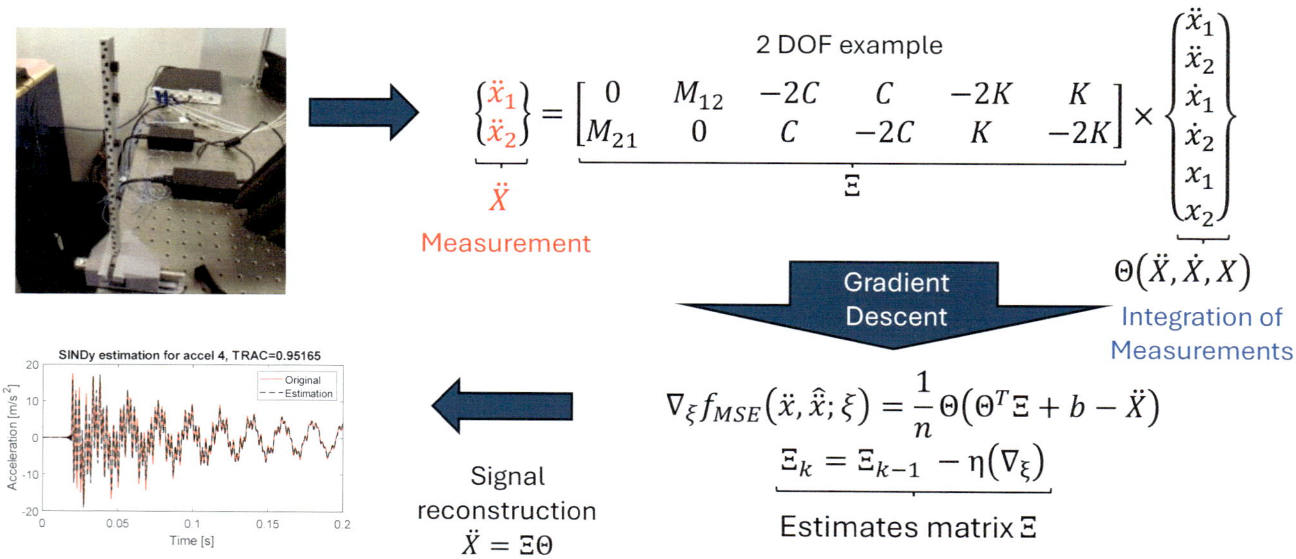

Fig. 1 Flowchart of proposed SINDy with stochastic gradient descent

$$\Xi_k = \Xi_{k-1} - \eta\left(\nabla_\xi\right) \tag{3}$$

where η is the learning rate and hyperparameter of the model. In addition, during the training, the values of the mass matrix which are equivalent to their respective results will be set to zero. Such that, it will prevent the optimization to obtain a trivial solution of $M_{ii}\ddot{X}_{ii} = M_{ii}\ddot{X}_{ii}$. In addition, considering that system identification is an ill-posed problem, with multiple possible solutions, extra information is required to obtain the correct system of equations. For instance, one of the physics terms added is related to the mass of the system. Multiple combinations of mass can be obtained for the same natural frequency and mode shape; therefore, the total mass of the system was required to obtain a more precise system of equations. From the total mass of the system ($\hat{M}$) it is assumed that the sum of the diagonal terms of the mass matrix should be the same value, as shown in Eq. (4).

$$\hat{M} = \sum_{i=1}^{n} M_{ii} \tag{4}$$

where, M_{ii} are the diagonal terms of the mass matrix and n is the number of DOF of the system. Later, the mass of the first DOF can be estimated based on the absolute error between the estimated mass and the total mass, as in Eq. (5).

$$M_{11}^{k+1} = M_{11}^{k} + \eta_m \left| \hat{M} - \sum_{i=1}^{n} M_{ii} \right| \tag{5}$$

Finally, assuming the parameters estimated on the optimization are normalized by the respective DOF mass, the other values of the mass matrix can be estimated based on Eq. (6).

$$M_{ii} = \frac{\xi_{i-1,i}}{\xi_{i.i-1}} \times M_{i-1,i-1} \tag{6}$$

where, ξ_{ij} is the term of the library at the ith row and j^{th} column. Another penalty term added to the optimization problem is based on the eigenvalues of the system. Since the estimated PDE can have different natural frequencies than the original system (based on the eigen solution), the mean squared error of the estimated system is compared to the measurements. To obtain a penalty term in the SGD the gradient of the eigenvalue with respect to the stiffness matrix is given by Eq. (7) [14, 15].

$$\frac{d\lambda}{dK} = U^T \otimes U^T \tag{7}$$

where, λ is one of the eigenvalues of the system, K is the stiffness matrix, and U is its mode shape. Then the stiffness matrix can be updated based on the gradient of the error of the eigenvalue as presented in Eq. (8).

$$\nabla f_{MSE}\left(\lambda, \widehat{\lambda}\right) = \frac{1}{\lambda}\frac{d\lambda}{dK}\left(\lambda - \widehat{\lambda}\right) \tag{8}$$

Afterwards, a term was added to the optimization related to the symmetry of the matrices. It is known that for linear vibrational systems, the mass, damping, and stiffness matrices are symmetric. To encourage symmetry in the estimation, the penalty term (A_p) was implemented as described in Eq. (9).

$$A_p = \frac{\left[A - A^T\right]}{\|A\|} \tag{9}$$

where A can be either of the parameters from the library, such as mass, stiffness or damping. Finally, the loss function of the optimization problem is given by a combination of the four terms described previously, as presented in Eq. (10).

$$\mathcal{L} = \nabla_\xi f_{MSE}\left(\ddot{x}, \widehat{\ddot{x}}, \xi\right) + \alpha_1 \nabla f_{MSE}\left(\lambda, \widehat{\lambda}\right) + \alpha_2 f_{MAE}\left(M, \hat{M}\right) + \alpha_3 A_p \tag{10}$$

where, α_i are the weights for each of the parameters and must be fine-tuned depending on the application.

Simulation Results

The proposed algorithm will be employed into a cantilever beam which will be measured by four accelerometers. In order to validate the results obtained, a comparison with a Finite Element Model (FEM) will be performed. The FE model consists of a one hundred element cantilever beam, made of aluminum with the following properties:

Table 1 Cantilever beam properties

Property	Young's modulus (psi)	Width (in)	Thickness (in)	Length (in)	Mass density
Value	10^6	0.75	0.25	9.5	2.5×10^{-4}

The model was created considering the displacement and rotation of the elements, leading to the following mode shapes:

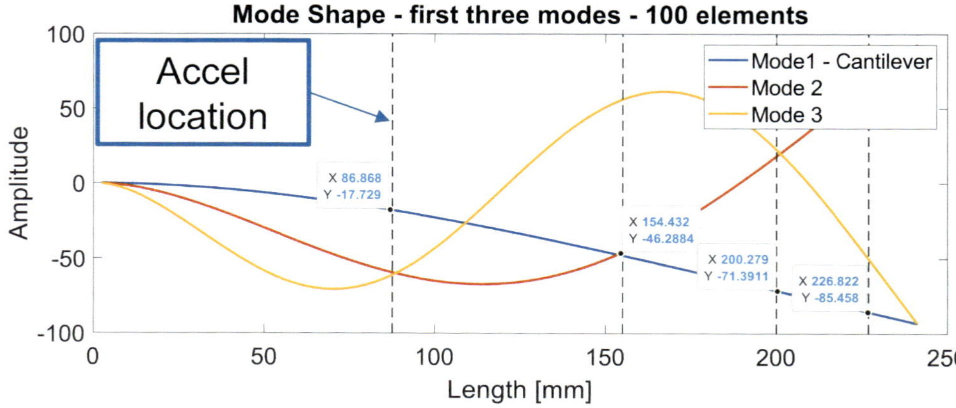

Fig. 2 Mode shapes obtained by FEM

After obtaining the mode shapes and natural frequencies from the simulation, the model was reduced into the same location of the accelerometers by using the System Equivalent Reduction Expansion Process (SEREP) [16]. The reduction of the model can be performed as shown in Eq. (11) below:

$$A_a^S = T_U^T A_n T_U \tag{11}$$

where A_a^S can be either the mass or stiffness matrix of the active set, A_n the matrix of the original model and T_U is the transformation matrix. Since the natural frequencies and mode shapes of the FE model differ from the measurements, a model update is required to represent the beam correctly. It was employed the stiffness and mass matrix analytical model improvement (KM-AMI) for the reduced model in order to obtain the correct governing equations for the system. The model improvement is calculated by Eqs. (12) and (13).

$$M_I = M_S + V^T \left[I - \overline{M}_S\right] V \tag{12}$$

where M_I is the improved mass matrix, M_S is the seed (or original) mass matrix, $\overline{M}_S$ is the orthogonal original mass matrix, however, instead of using the mode shapes obtained from the simulation, the measured mode shapes. Finally, V is a transformation matrix obtained from the measured mode shapes, the original mass matrix and $\overline{M}_S$.

$$K_I = K_S + V^T \left[\Omega^2 + \overline{K}_S\right] V - [K_S U_I V] - [K_S U_I V]^T \tag{13}$$

where similarly to the equation of the mass matrix, K_I, K_S and $\overline{K}_S$ are the improved and original stiffness matrices and the orthogonal stiffness matrix respectively. In addition, Ω^2 are the eigenvalues of the FEM and U_I its mode shapes. After performing the model reduction and updating it, its mode shapes are presented in Fig. 3. Since only three accelerometers were used for the SINDy estimation, the updated mode shapes only have three points, thus the plots do not look as smooth as the original FE simulation. It can be seen that the updated model and experimental results align very well, and in addition the natural frequencies of the mode shapes are the same for experimental and simulation.

The stiffness and mass matrices obtained after the model update will then be used as a metric to evaluate the parameters obtained by the SINDy algorithm. The values obtained from the FE model can be seen in Eq. (14) below:

$$M_{FEM} = \begin{bmatrix} 0.7 & -0.5 & 0.6 \\ -0.5 & 1.3 & -1.1 \\ 0.6 & -1.1 & 1.7 \end{bmatrix} \text{ and } K_{FEM} = \begin{bmatrix} 4.9 & -4.7 & 1.9 \\ -4.7 & 6.0 & -2.6 \\ 1.9 & -2.6 & 1.2 \end{bmatrix} 10^7 \tag{14}$$

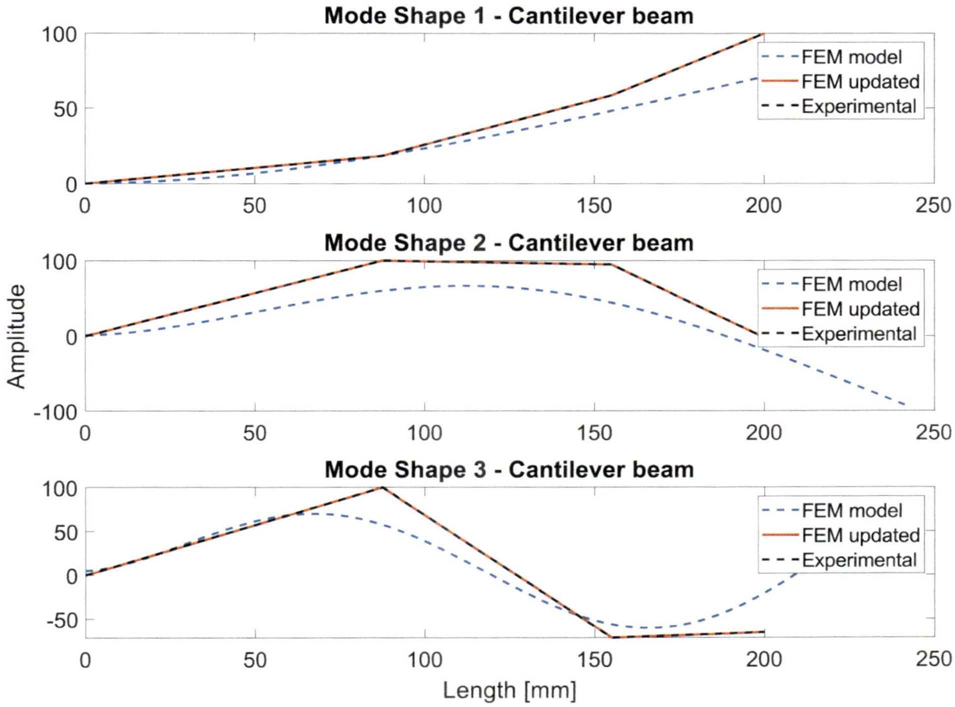

Fig. 3 Comparison of original simulation, updated model, and experimental mode shapes

Learning Results and Discussion

As a benchmark of the proposed methodology and the FEM simulation, the results presented next will be a comparison between the original approach for system identification and the physics-informed approach. Initially, the system identification using SINDy can be seen in Fig. 4.

As presented in the figure, the approximation of the first mode shape has a high accuracy, presenting 100% of modal assurance criteria (MAC) for the third and 99% for the fourth of the SINDy modes. However, when looking at the second mode shape, only the first of the estimated modes has good accuracy, with 75% MAC. When looking at the time reconstruction of the measurement it can be seen that for the accelerometers located in places where the first mode shape is predominant the accuracy of the time reconstruction is much higher, achieving above 90% of the time reconstruction assurance criteria

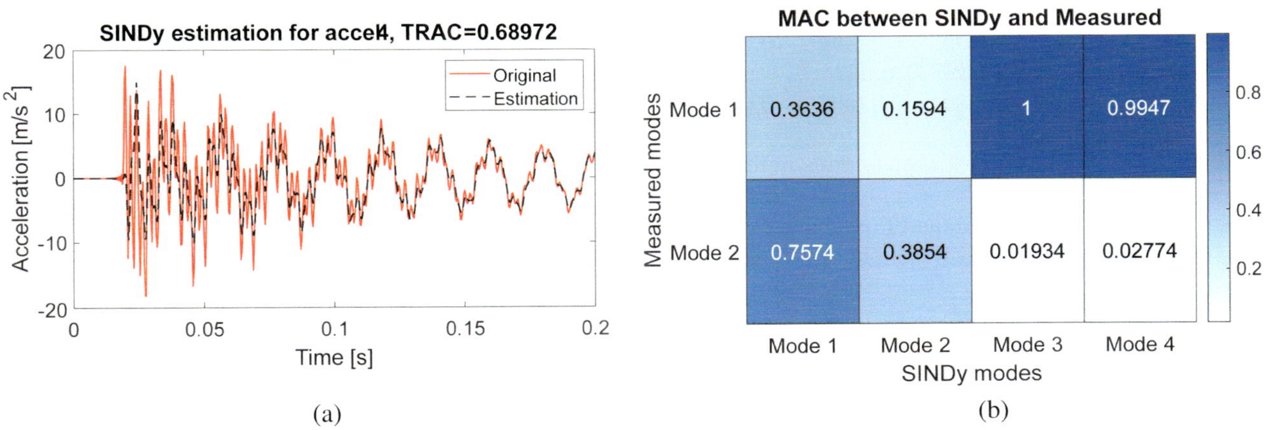

Fig. 4 Result from original SINDy, (a) Signal reconstruction of accelerometer 4 and (b) MAC comparison with measurements

(TRAC). Meanwhile, for accelerometers such as accel 4, which is located in a region where higher frequency is predominant, the accuracy for TRAC is only 68%. Since the proposed approach uses an SGD approach, there are hyperparameters to be selected that will influence the final reconstruction, Table 2 shows the values selected for the hyperparameters.

Table 2 Hyperparameters for SGD SINDy training

Hyperparameter	**# of Epochs**	**Batch Size**	**Learning Rate**	**Mass LR**	α_{Freq}	$\alpha_{Symmetry}$
Value	10,000	128	1.5×10^{-4}	1.5×10^{-4}	0.1	1

After applying the same dataset into the proposed approach, the results are presented in Fig. 5.

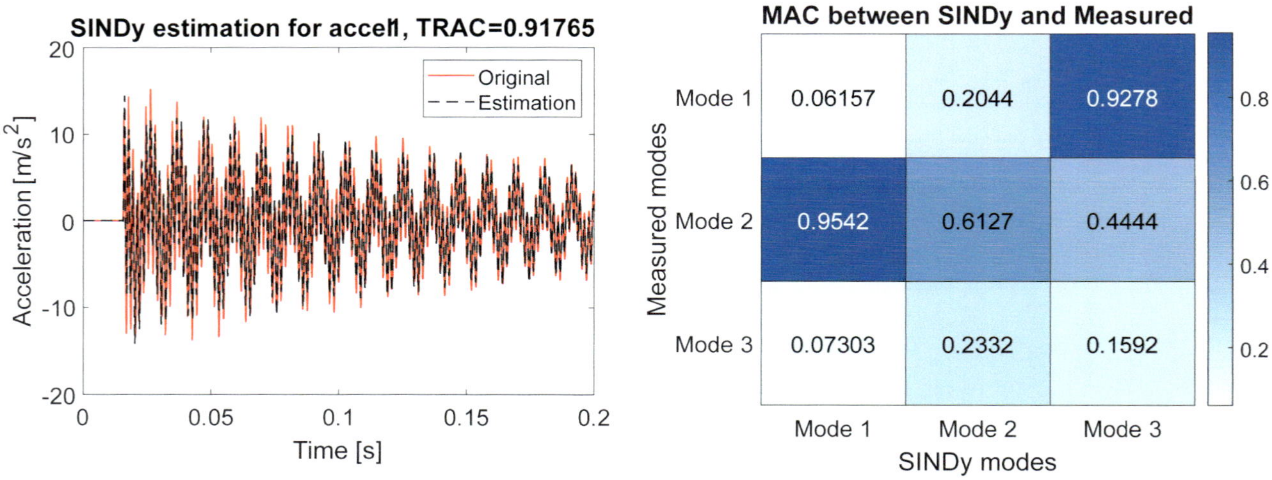

(a) (b)

Fig. 5 Results from SGD SINDy with penalty terms, (a) Signal reconstruction of accelerometer 1 and (b) MAC comparison with measurements

As presented in Fig. 5a, the proposed technique has a higher accuracy for accelerometers closer to the base of the beam, which will have higher frequency content available. For this case, a TRAC of 91% is obtained, furthermore, all other accelerometers have an accuracy above 90%. In addition, as presented in Fig. 5b, the mode shape reconstruction has higher than 90% MAC for both mode shapes 1 and 2. Unfortunately, due to the numerical damping caused by the integration of the acceleration, the third mode shape could not be obtained. In addition to the signal reconstruction and MAC, the matrices are compared between a finite element (FE) model and the SINDy estimation. The mass and stiffness matrices for SINDy are presented in Eq. (15) below.

$$M_{SINDy} = \begin{bmatrix} 0.0409 & -0.0395 & 0.0086 \\ -0.0395 & 0.0425 & -0.0058 \\ 0.0080 & -0.0058 & 0.0165 \end{bmatrix} \text{ and } K_{SINDy} = \begin{bmatrix} -1.825 & 0.764 & -2.039 \\ 0.753 & -1.683 & 5.678 \\ -2.064 & 5.658 & 1.733 \end{bmatrix} 10^3 \tag{15}$$

As shown in Eq. (15), the off-diagonal terms in the matrices are similar, differing only two digits after the decimal for the stiffness matrix. Leading to a behavior that is expected for a linear vibrational system. In addition, the difference between the values obtained by SINDy and the FE model is computed, following Eq. (16), where the absolute error is calculated based on the normalized values of each matrix.

$$Error_A = \frac{[A]_{FEM}}{\|[A]_{FEM}\|} - \frac{[A]_{SINDy}}{\|[A]_{SINDy}\|} \tag{16}$$

Figure 6 presents the computed error between the FE model and the SINDy estimation for the mass matrix (a) and the stiffness matrix (b). It can be noticed that although the SINDy model obtains a symmetric result for the system for both matrices, the proportions of the matrix are different when compared to the FE model. For instance, the highest value of the FE model stiffness matrix is on the last DOF, while for the SINDy estimation, it happens in the first DOF.

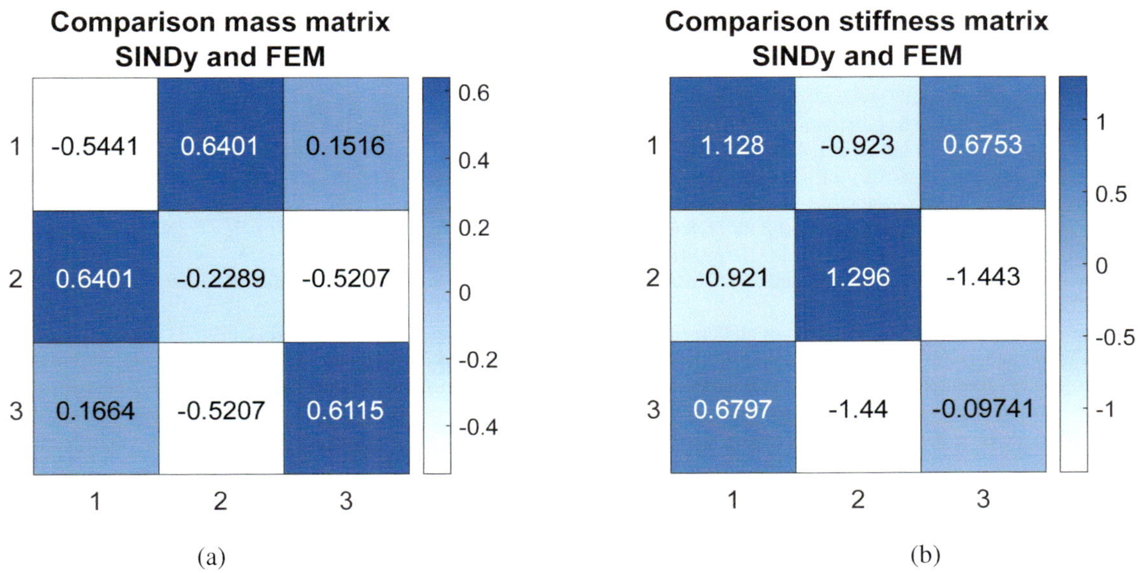

Fig. 6 Comparison of (a) mass and (b) stiffness matrices between FE model and SINDy estimation

Conclusion

This study presents a variation of the SINDy algorithm developed by Brunton et al. [5], for application on second order PDEs, specifically vibrational systems. Results for original SINDy show acceptable approximation for governing equation estimation, however, due to the dependence between variables in the problem, it becomes challenging to correctly predict mode shapes and natural frequencies of the structures. The modifications in the algorithm allow a better estimation of both natural frequencies and mode shapes, while increasing the signal reconstruction obtained by the estimated parameters. While the PDEs estimated using SINDy are not the same as the ones obtained by a simplified FE model, the proposed approach suggests a good approximation of the system, with a small number of assumptions that can be easily obtained by measurements.

For future work, more improvements for the algorithm can be performed, for instance the current version works best for free vibration measurements (obtained by impact test). When different excitation is given, e.g. random excitation, the governing equations identification poses to be challenging. In addition, more tests can be performed in more complex structures, including nonlinear structures to observe how the approach will behave when dealing with more parameters to be estimated.

Acknowledgments This work is supported by the Department of Energy Kansas City National Security Campus Managed by Honeywell under contract # N000510678. NSC-614-5662 10/2024. Unclassified Unlimited Release

References

1. O. Nelles and O. Nelles, *Nonlinear dynamic system identification*. Springer, 2020.
2. F. Abdullah and P. D. Christofides, "Data-based modeling and control of nonlinear process systems using sparse identification: An overview of recent results," *Computers & Chemical Engineering*, vol. 174, p. 108247, 2023.
3. T. Liu and H. Meidani, "Physics-informed neural networks for system identification of structural systems with a multiphysics damping model," *Journal of Engineering Mechanics*, vol. 149, no. 10, p. 04023079, 2023.
4. M. Raissi, P. Perdikaris, and G. E. Karniadakis, "Physics-informed neural networks: A deep learning framework for solving forward and inverse problems involving nonlinear partial differential equations," *Journal of Computational physics*, vol. 378, pp. 686-707, 2019.
5. S. L. Brunton, J. L. Proctor, and J. N. Kutz, "Discovering governing equations from data by sparse identification of nonlinear dynamical systems," *Proceedings of the national academy of sciences*, vol. 113, no. 15, pp. 3932-3937, 2016.
6. C. Lathourakis and A. Cicirello, "Physics-enhanced sparse identification of nonlinear oscillator with coulomb friction," in *International Conference on Nonlinear Dynamics and Applications*, 2023: Springer, pp. 507-517.
7. C. Lathourakis and A. Cicirello, "Physics enhanced sparse identification of dynamical systems with discontinuous nonlinearities," *Nonlinear Dynamics*, pp. 1-28, 2024.

8. S. Piramoon, M. Ayoubi, and S. Bashash, “Modeling and Vibration Suppression of Rotating Machines Using the Sparse Identification of Nonlinear Dynamics and Terminal Sliding Mode Control,” *IEEE Access*, 2024.
9. S.-C. Huang, H.-W. Chen, and M.-H. Tien, “Predicting Nonlinear Modal Properties by Measuring Free Vibration Responses,” *Journal of Computational and Nonlinear Dynamics*, vol. 18, no. 4, p. 041005, 2023.
10. K. Kaheman, J. N. Kutz, and S. L. Brunton, “SINDy-PI: a robust algorithm for parallel implicit sparse identification of nonlinear dynamics,” *Proceedings of the Royal Society A*, vol. 476, no. 2242, p. 20200279, 2020.
11. P. Goyal and P. Benner, “Discovery of nonlinear dynamical systems using a Runge–Kutta inspired dictionary-based sparse regression approach,” *Proceedings of the Royal Society A*, vol. 478, no. 2262, p. 20210883, 2022.
12. Y. Liu, S. Shi, Y. Fang, and A. Fu, “Robust data-driven wave excitation force estimation for wave energy converters with nonlinear probabilistic modelling,” *Ocean Engineering*, vol. 310, p. 118726, 2024.
13. S. H. Rudy, S. L. Brunton, J. L. Proctor, and J. N. Kutz, “Data-driven discovery of partial differential equations,” *Science advances*, vol. 3, no. 4, p. e1602614, 2017.
14. J. R. Magnus, “On differentiating eigenvalues and eigenvectors,” *Econometric theory*, vol. 1, no. 2, pp. 179-191, 1985.
15. A. D. Belegundu and T. R. Chandrupatla, *Optimization concepts and applications in engineering*. Cambridge University Press, 2019.
16. R. Allemang and P. Avitabile, *Handbook of experimental structural dynamics*. Springer Nature, 2022.

Chapter 5
Preliminary Investigation of an Operating Wind Turbine in a Seismic-prone Area

Mona Amer, Tomás Núñez, and Carlos E. Ventura

Abstract Areas with high wind resources for wind turbines often coincide with earthquake risk areas. To enlarge the application of wind turbines in these regions, it is thus, necessary to study and understand the dynamic behavior of operating wind turbines in seismic areas. Currently, there is a lack of experimental studies on these types of structures under seismic conditions. The proposed research shall change the state of the art by performing a long-term field study on wind turbines located in high-risk earthquake regions. In a period of at least one year, data will be continuously captured. In a high-risk earthquake country, such as Chile, an earthquake with a magnitude of six or higher occurs in almost every year. The data will be used to draw conclusions on the dynamic behavior of wind turbines under seismically induced loads and to derive suitable structural health monitoring techniques. For a preliminary analysis, data from a 20-hour-measurement of a wind turbine in Chile is used. Challenges associated with the data analysis are identified and discussed to prepare the following long-term field study. These insights of the proposed study will enhance the development of structural health monitoring serving as part of digital twins for wind turbines.

Keywords Wind energy · Earthquake Engineering · Structural Health Monitoring · Operational Modal Analysis · Digital Twin

Introduction

The transition to renewable energy is one of the key challenges faced by our current society. Wind energy represents one of the main pillars for reaching the energy transition. Areas with high wind resources often coincide with regions that have a high seismic activity [1]. For this reason, it is important to study and comprehend the structural behavior of wind turbines in these areas and under seismically induced loads to adjust the design guidelines and develop suitable structural health monitoring techniques.

Currently, the structural design of wind turbines is not standardized within the building code and design guidelines only rely on simulations in the case of earthquakes [2]. For this reason, there is no consistent procedure on how to perform structural analysis and mitigate structural failure in the design process. Simulations are subjected to simplifications some of which are not tenable in case of earthquakes. Under earthquake loads usually negligible effects gain more importance [2], such as the influence of higher modes, since the vertical modes (usually associated with higher natural frequencies) are excited during an earthquake [3]. This, in turn, can have an impact on maximum deflections at the top of the tower, which can also affect rotating components in the nacelle and can lead to potential damage. The only published experimental study has been performed by Prowell et al. [4] on a scaled wind turbine that has been excited by an outside shake table. A long-term field study actually capturing the load of earthquakes acting on the wind turbine structure is necessary to generate important validation data. This applies especially to specific parameters such as the damping behavior, which is usually determined experimentally. The experimental validation data will provide consistent guidelines for the structural design and assessment of wind turbines in seismic areas. Therefore, such a study is proposed and described in detail below.

Mona Amer · Tomás Núñez · Carlos E. Ventura
Department of Civil Engineering, The University of British Columbia, Vancouver, BC V6T 1Z4
e-mail: mona.amer@ubc.ca; nuneztom@mail.ubc.ca; ventura@civil.ubc.ca

Brandon Dilworth et al. (eds.), *Topics in Modal Analysis & Parameter Identification II, Vol. 10*, Conference Proceedings of the Society for Experimental Mechanics Series, https://doi.org/10.13052/9788743801559-5

Methodology

For the realization of the proposed study, an earthquake-intensive country such as Chile is chosen, where almost every year an earthquake with a magnitude greater than six is recorded, causing damage to buildings and other structures. Up to four wind turbines will permanently be instrumented with acceleration sensors, which will transmit the acquired data to a remote server, to which researchers have access. For the measurement, senors of type ADXL 355 can be applied, which have proven successful in previous structural health monitoring approaches and represent a cost-effective solution at the same time [5]. Data is recorded over a long period until a significant earthquake event can be recognized. After the earthquake event, another field trip will be scheduled to assess the structural health of the wind turbine that has been seismically excited. This knowledge and experimental data set will help understand, what effects play a major role during an earthquake and what measures have to be taken by potential structural health monitoring approaches to avoid structural damage.

Since there is no data from a real wind turbine under seismic excitation available yet, data from a 20-hour-measurement of a wind turbine in the South of Chile close to Los Ángeles is used for preliminary investigations. The turbine is rotating during measurements. Understanding the dynamic behavior of a wind turbine under similar conditions as the ones studied at a later point in time, will help to set up a suitable test plan and to identify the challenges associated with the data analysis.

The modal identification of modal parameters in operation is the core objective of the data analysis. Modal parameters can be tracked over time. An observed shift in these parameters can be evidence of structural changes and allows for early on maintenance before any damage occurs [6]. This knowledge can thus be used to design structural health monitoring approaches. The development of a dynamic model by means of modal parameters is the focal point of the ongoing evaluation.

The structural information is sampled with a frequency of 128 Hz by means of four velocity sensors, which are placed inside the tower. Two sensors are opposite to each other on top of the tower, just below the nacelle, and the other two sensors are distributed along the height of the tower.

It is important to mention that the rotational speed signal is sampled at $f_s = 0.0017$ Hz, which means that there is one time stamp every 10 minutes capturing this dimension. The rotational speed is depicted as a function of time during the whole measurement in Fig. 1. For the analysis, time sections are chosen, where the speed remains constant for at least two consecutive time stamps. Three time ranges with an average constant rotational speed of 6 rpm, 9 rpm, and 10.5 rpm are chosen for the preliminary analysis. These sections are marked in Fig. 1 with red color. Nevertheless, speed fluctuations within the selected data sets can occur the extent of which remains unknown.

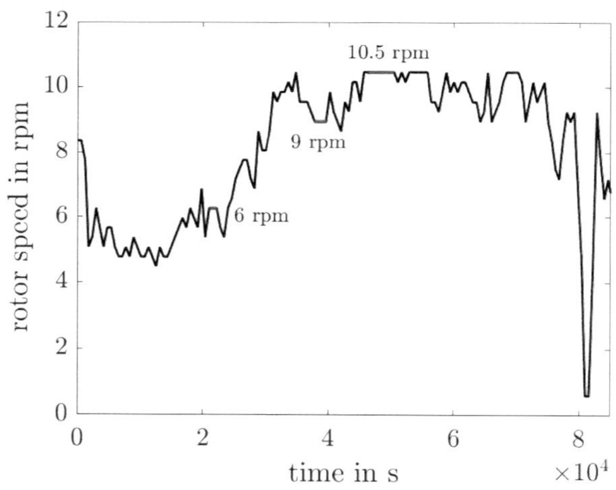

Fig. 1 Rotational speed as a function of time during the ambient vibration tests.

Results

In the following section, the challenges arising from modal identification are discussed. The modal identification from ambient vibration tests of a structure with rotating components is always associated with the problem of superimposing harmonic response parts that hinder the reliable identification of structural parameters.

From standstill conditions, it is known that the lowest natural frequencies are expected around 0.14 Hz. The next natural frequencies are at 0.56 Hz, 0.67 Hz, 1.1 Hz, 1.16 Hz, and 1.54 Hz. Subsequently, the data of the highest rotational speed at 10.5 rpm will be discussed in more detail. The first rotor harmonic is at approximately 0.175 Hz. Due to the number of blades, the third harmonic is also significant in the measured response data. Beyond this, the influence of the sixth harmonic can be observed. It is evident that the frequencies of the harmonics are very close to some of the structural frequencies or even overlap with them completely. A strong influence of the rotation in the data is expected.

In [7], a technique was proposed to eliminate the harmonic response from the measured response and obtain the purely stochastic part of the vibration response. This, in turn, can be used as input to established algorithms for operational modal analysis. For the technique, a rotary encoder signal is used, which provides the phase information of the rotor at every measured time instant. This allows for a minimally-invasive procedure with respect to the measured structural response and can additionally account for rotor speed fluctuations.

Since the rotary encoder signal is not available, the approach for the elimination of the harmonic signal content presented in [7] cannot be applied. A phase signal of the rotor synchronized with the actual structural response measurement would have been necessary. Instead, the approach described in [8] is applied to the data. The rotational speed is estimated as one constant value over the measurement period.

In Fig. 2, the stability diagram resulting from the modal analysis is illustrated over the first three singular values (SVs) of the resulting power spectral density (PSD) before and after the elimination of the harmonic response. In this case, the elimination leads to a strong effect on the neighboring frequency area that results in the accumulation of calculated poles in the area next to the frequency range influenced by the elimination. Adjusting the different parameters used within this techniques could lead to an improved result. The preliminary analysis shows the dominance of the harmonic effect, which is superimposing the structural natural frequencies of the system that cannot be identified reliably. The result also shows that within the whole measurement period of 80 minutes, strong rotor speed fluctuations occur, leading to the wide peak of the harmonic in the frequency domain and the strong influence when eliminating the harmonic response part. The preliminary analysis shows that there is a need for application-specific techniques to deal with the harmonic effect in the data. Testing different techniques and developing suitable methods to handle the harmonic influence is part of future research.

Another approach to identify harmonics, is the use of the probability density function (PDF). According to the shape of the PDF belonging to the mode investigated, a user can classify this mode as a structural mode, a harmonic, or even a mathematical pole that has been calculated, but does not represent any physical behavior, compare Fig. 3a. An example, for the third calculated mode in operation at around 1 Hz is depicted in Fig. 3b. From Fig. 3 it can be concluded that the interpretation of the PDF is complicated, since the PDF does not show a clear shape, which does belong to one of the defined possibilities. The PDF depicted in Fig. 3b rather describes a mixture of the possible scenarios. This underlines, once more, the complexity of the available data set from the rotating wind turbine and the necessity for specific tools to reliably identify the modal parameters.

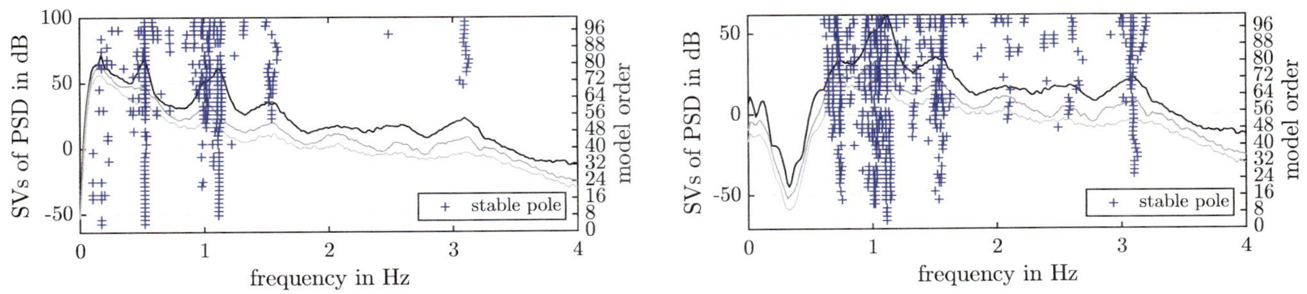

(a) before harmonic elimination

(b) after harmonic elimination

Fig. 2 Stability diagram displaying the first three singular values of the measured structural response's PSD before and after the elimination of the first harmonic.

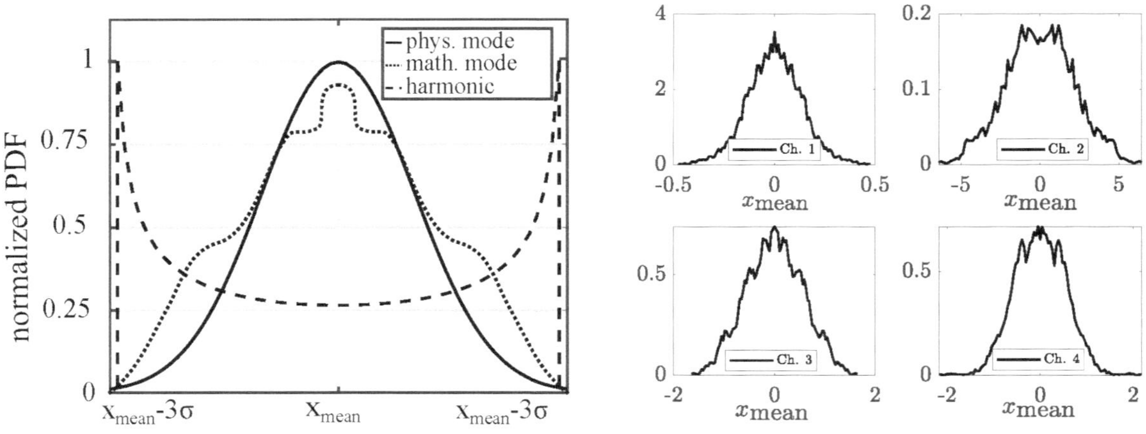

(a) Example of PDFs (b) PDF of calculated mode

Fig. 3 Probability density function (a) depicted as examples for the different types of modes, (b) of each measurement channel for a calculated mode at approx. 1 Hz.

Conclusions

For a better understanding of the dynamic behavior of wind turbines in earthquake regions, a long-term field study, instrumenting several wind turbines in high-risk seismic areas, is intended. To study the characteristics, of wind turbines in such kind of areas, a 20-hours measurement from a wind turbine operated in the South of Chile served as a basis for a preliminary analysis. From this analysis, it can be concluded that the influence of the rotation is predominant and suitable measures have to be found and applied to the data to enable the reliable modal identification. The only way for a successful data analysis, is the elimination of the harmonic response part, since the harmonic frequencies are too close to the natural frequencies of the system. An estimation of the harmonic response using a Kalman filter approach in combination with the row space projection of the raw data onto the orthogonal complement of the row space of the estimated periodic part as proposed in [9] could serve as a way to deal with the strong harmonic influence. Based on these future results, further conclusions can be drawn and necessary adjustments can be made to find a suitable technique to treat the harmonics and be able to reliably identify the modal parameters in operation. Having identified a modal model under varying load conditions, especially during ground motion excitation, will enable the development of structural health monitoring approaches that can serve as part of digital twins for wind turbines in seismic areas.

References

1. Altunsu, E., Gunes, O., and Sarı, A. "Design principles for wind turbine earthquake and wind load combinations". *International Journal of Steel Structures*, 22(3):791–804 (2022).
2. Katsanos, E.I., Thöns, S., and Georgakis, C.T. "Wind turbines and seismic hazard: a state-of-the-art review". *Wind Energy*, 19(11):2113–2133 (2016).
3. Kaynia, A.M. "Seismic considerations in design of offshore wind turbines". *Soil Dynamics and Earthquake Engineering*, 124:399–407 (2019).
4. Prowell, I., Veletzos, M., Elgamal, A., and Restrepo, J. "Experimental and numerical seismic response of a 65 kw wind turbine". *Journal of Earthquake Engineering*, 13(8):1172–1190 (2009).
5. Crognale, M., Rinaldi, C., Potenza, F., Gattulli, V., Colarieti, A., and Franchi, F. "Developing and testing high-performance SHM sensors mounting low-noise MEMS accelerometers". *Sensors*, 24(8):2435 (2024).
6. Goyal, D. and Pabla, B. "The vibration monitoring methods and signal processing techniques for structural health monitoring: a review". *Archives of Computational Methods in Engineering*, 23:585–594 (2016).
7. Amer, M., Ventura, C.E., Maroldt, N., Seume, J.R., and Wallaschek, J. "Modal parameter estimation of turbomachinery in operation taking into account friction damping". *Mechanical Systems and Signal Processing*, 216:111414 (2024).
8. Zivanovic, M., Plaza, A., Iriarte, X., and Carlosena, A. "Instantaneous amplitude and phase signal modeling for harmonic removal in wind turbines". *Mechanical Systems and Signal Processing*, 189:110095 (2023).
9. Greś, S., Döhler, M., Andersen, P., and Mevel, L. "Kalman filter-based subspace identification for operational modal analysis under unmeasured periodic excitation". *Mechanical Systems and Signal Processing*, 146:106996 (2021).

Chapter 6
Modal Property Identification of the Bear Mountain Bridge

CDT Gautam Namjoshi, CDT Philip Williams, CDT Ethan Espinosa, and Dr. Thomas Matarazzo

Abstract The Bear Mountain Bridge is a historic suspension bridge located in the lower Hudson Valley of New York. A sensor network consisting of eight accelerometers installed on the bridge's main suspended span is used to collect operational vibrations for modal property identification. This paper analyzes the data recorded by the fixed sensor network to estimate the bridge's vertical and torsional modal frequencies and mode shapes. Three different system identification methods were implemented to estimate modal properties. Through direct comparisons of the estimated frequencies and stabilization diagrams, modal frequencies and mode shapes were evaluated. Overall, thirteen possible modal frequencies were identified below 2.5 Hz, of which six modes were cross validated by all three methods. This paper reports the first known estimates of the modal properties of this bridge.

Keywords System Identification · Bridge · Modal Properties · Sensor Network · Structural Health Monitoring

Introduction

Bridge inspections can be a costly and time-intensive process that generally does not evaluate structural dynamics behavior. Structural dynamic properties are important as they are linked to the mass and stiffness of the system. Over the past few decades, numerous vibration-based structural health monitoring methods have been developed with the goal of determining dynamics properties and inferring structural condition. Ambient modal analysis is the practice of recording structural vibrations in operational conditions and using these measurements to estimate structural modal properties [1]. Such dynamic properties are useful for calibrating finite-element models, computing damage-sensitive features, supporting condition assessments, and other applications [2].

The Bear Mountain Bridge was opened in 1924 and possesses cultural and economic significance for Hudson Valley region of New York State [3]. As a suspension bridge, its modal properties are generally comparable to other cable-supported, long-span bridges, such as the Golden Gate Bridge, on which similar characterizations have been conducted [3, 4]. In this study, vertical and torsional modal frequencies and mode shapes of the Bear Mountain Bridge are estimated from the accelerometer data. The determination of modal properties of the bridge supports a baseline, from which, changes may be more easily identified. Such a baseline should be calibrated with respect to environmental variables to better infer a change in structural state.

Data Collection

Modal identification in this study is based on the measurements of a network of accelerometers installed on the bridge. The accelerometer sensor network is comprised of eight accelerometers with two on the North side of the bridge, and six on the South side of the bridge (see Fig. 1). Sensors are distributed across the 1632 feet long main span of the bridge, to capture spatial dynamics of the suspended main span [3]. The New York State Bridge Authority personnel greatly assisted with the planning and installation of the sensor network. All accelerometers were hard-wired to a base station (data acquisition system) on the West side of the bridge. Sensors were affixed to the bridge using a combination of mounting adhesive and clamps and positioned in such a way to resist detrimental environmental conditions.

CDT Gautam Namjoshi · CDT Philip Williams · CDT Ethan Espinosa · Dr. Thomas Matarazzo
Department of Civil and Mechanical Engineering, United States Military Academy, West Point, NY 10997
e-mail: philip.williams@westpoint.edu; gautam.namjoshi@westpoint.edu; ethan.espinosa@westpoint.edu; thomas.matarazzo@westpoint.edu

Brandon Dilworth et al. (eds.), *Topics in Modal Analysis & Parameter Identification II, Vol. 10*, Conference Proceedings of the Society for Experimental Mechanics Series, https://doi.org/10.13052/9788743801559-6

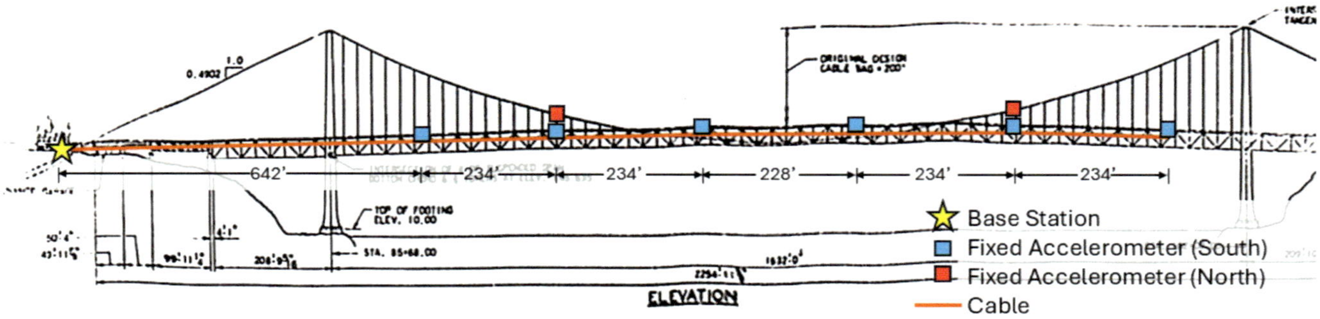

Fig. 1 Overview of Sensor Network Layout

Data collected from the "fixed" set of sensors requires on-site personnel to collect data, and so the data set is restricted to the bridge's working hours of 0800-1600. Data was collected in 2022 between the months of May and August. The sensors are effective at all ranges, including the expected range of 0-2.5 Hz for the relevant modal frequencies of the bridge.

Modal Property Identification Methods

The principal challenge with determining the modal properties of the Bear Mountain Bridge stems from the quality of the bridge response data. Since this study considers operational modal analysis data, output-only system identification methods are most appropriate. Three different output-only system identification methods were used to analyze the operational acceleration data: Automated Frequency Domain Decomposition (AFDD) [5], Eigensystem Realization Algorithm-Natural Excitation Techniques (ERA-NExT) [6], Covariance-driven Stochastic Subspace Identification(SSI-COV) [7]. These three output-only methods formed the primary methods for analysis; results from these methods were then compared within datasets via stabilization diagrams to identify stable modes and between datasets using histograms to identify the occurrences of modal frequencies. In addition to the lack of input data, due to technical difficulties with the performance of the sensor network, some erroneous datasets were discarded.

Determination of Reference Modal Frequencies

To determine reference frequencies based on their prevalence in the results, AFDD methods and ERA-NExT methods—across a range of model orders from ten to sixty—were used as they are quick at identifying modal properties. AFDD has proven to be an effective and computational efficient method to approach capable of finding closely spaced modes [5]. ERA-NExT is a modification of the Eigensystem Realization Algorithm that embeds an assumption for random, uncorrelated input excitation, which is best suited for ambient vibration [6, 8]. While ERA-NExT has the capability to generate stabilization diagrams, for the purposes of this paper only the generated frequencies were used as results were more accessible.

These methods were primarily used to generate reference modal frequency estimates. The computational efficiency of these two methods allowed for execution of the methods across a large number of datasets, allowing for a histogram of results to be generated as seen in Fig. 2. Should the frequency generated be a likely real modal frequency, it should be time-invariant and therefore be present in multiple datasets. The peaks of the histogram should then represent accurate estimates of the modal frequencies of the bridge. In order to determine peak selection, a statistical analysis of the results through a cumulative distribution function was conducted. Occurrence counts over $n = 583$ contained the top 20% of data, therefore this value was selected as a minimum value for selecting peaks as reference frequencies. While not a detailed stabilization diagram analysis, this method of data analysis is scalable and efficient, allowing for accurate reference estimates to then be compared against other methods' stabilization diagrams.

Stabilization Diagrams through SSI-COV

To further validate the found frequencies, stabilization diagrams were developed through one additional method, SSI-COV. Results from the diagram were then cross validated with the reference frequencies generated by the histogram. SSI-COV was initially developed to simplify the computation of SSI, utilizing weighted data correlations instead of orthogonal projections

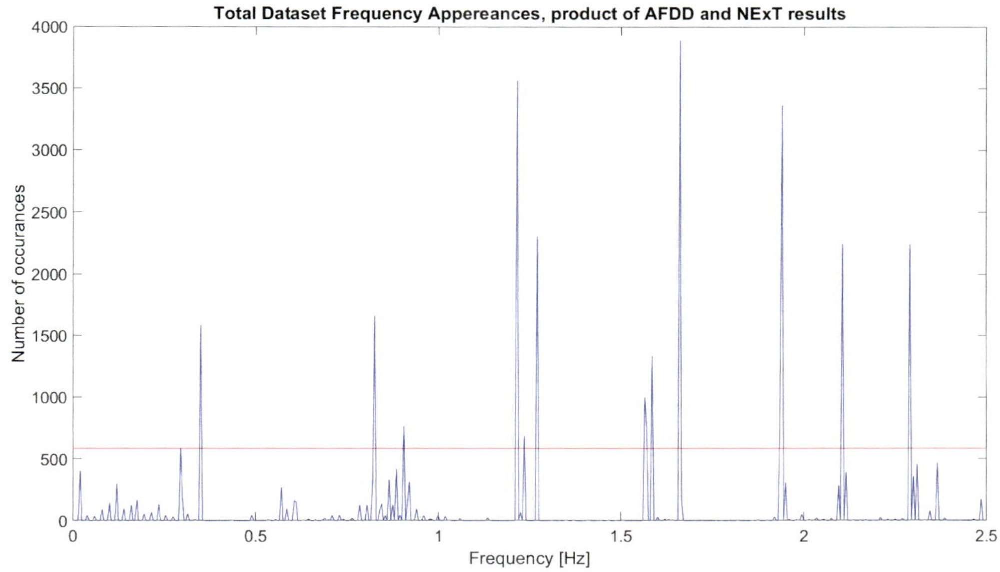

CDF Cutoff Value n=583 indicated by red line

Fig. 2 Histogram of AFDD/ERA-NExT (considering all 393 data sets)

to determine stable modal properties [7]. The tangible benefit of using this model, however, is the ease of developing stabilization diagrams. Stabilization diagrams help verify modal properties within the same dataset by finding patterns of stability. Vertical lines that appear on the graph indicate a trend of stability as modal order increases, as the continued presence of stable poles as the complexity of the model increases is indicative of real, stable modal frequencies. Selecting stable modes through this action helps distinguish likely modal frequencies against computational artifacts. As seen in Fig. 3, vertical lines are visible on the stabilization diagrams, identifying those stable modal frequencies. Limitations caused by the data

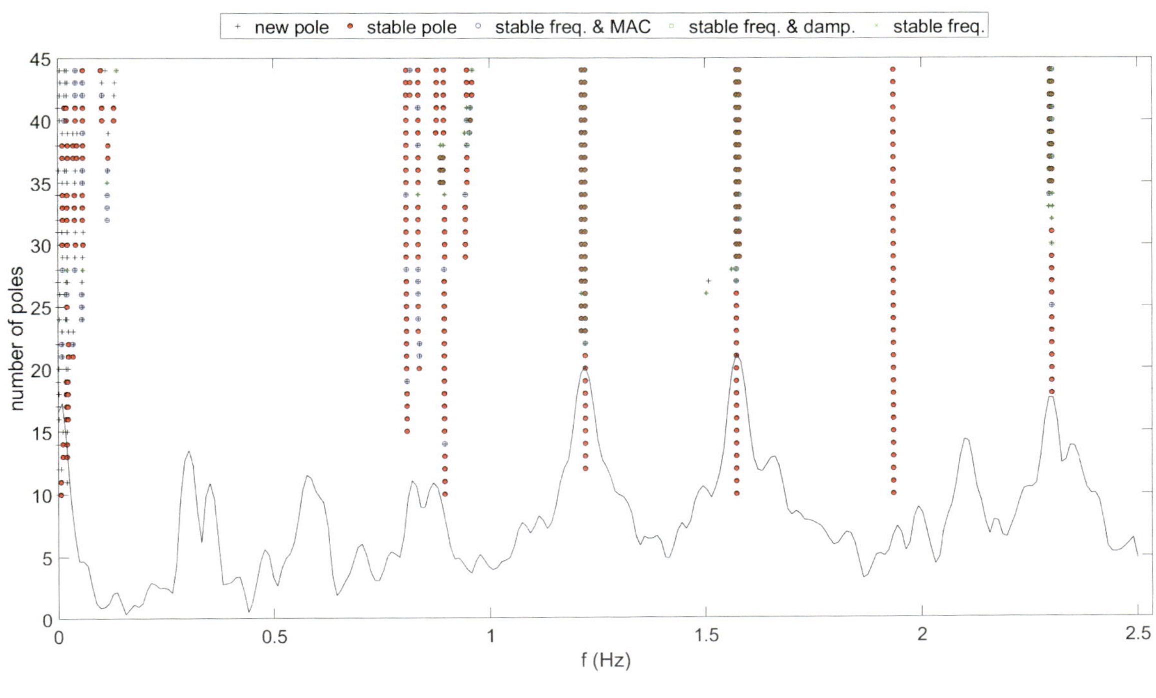

Fig. 3 Sample Stabilization Diagram for SSI-COV (one dataset)

collection method are also visible here. Due to the lack of precision of the fixed sensors at low frequencies, the stability for frequencies below 1 Hz is not as apparent as those above 1 Hz.

The diagram of stable modal frequencies as generated in Fig. 3 helps identify those specific frequencies for further analysis. Stable modes are considerably more useful for analysis, as the process of selecting stable frequencies removes computational artifacts present in other output-only models, refining the list of results and reducing the workload of analyzing mode shapes since instead of investigating all frequencies, thorough investigation can be limited to stable frequencies.

Results

The histogram in Fig. 2, gives us a foundational understanding of the spread of frequencies found by the AFDD and ERA-NExT methods across various datasets. The frequencies that make up the peaks provide us with a set of reference frequencies that are more likely to be modal frequencies due to their prevalence within the full set of data. In addition, these frequencies are validated by comparing them to the results of the stabilization diagrams from the SSI-COV model. While SSI-COV utilizes results from comparing different model orders within the *same* dataset the AFDD/NExT scaled results use data from *multiple* datasets from the same day. By validating results from both within datasets and between datasets, frequencies that appear across these different methods gain reasonable certainty in representing real modal frequencies.

Analyses of the mode shapes is simplified once the modal frequencies are known. By comparing the mode-shapes of different models at similar frequencies, as seen in Fig. 4, it is clear that method type, while impacting the exact values of the mode shape, does not matter when analyzing the general shape of the mode as the three methods produce visually similar mode shapes for similar modal frequencies. The visual correlation between these mode shapes serves as evidence towards the likelihood of these modal frequencies being real modal frequences, as similar mode shapes would not be found if the results were computational artifacts. Another critical component of the mode shape analysis is the type of mode found. Having sensors placed on both sides of the bridge allows for a determination of one of two modes. Torsional modes are indicated by opposing sensor movements for either side while vertical modes are indicated by unison sensor movements. Torsional Modes can be found in Mode 4 in Fig. 4, where sensors on the north side of the bridge are clearly moving opposite the motion of the south side. Vertical modes make up the rest of the stable modes identified in Table 1 and depicted in Fig. 4. Analysis of the type of mode, seen in Table 1, provides valuable information on the properties of the bridge under those specific frequencies, allowing for a better baseline understanding of the modal properties of the structure.

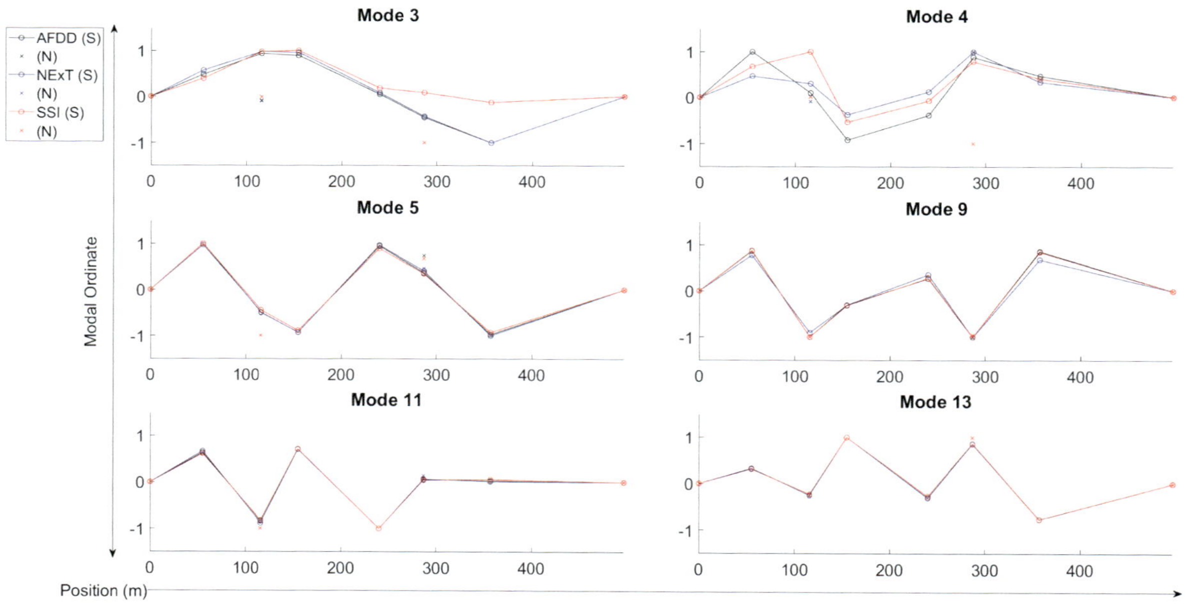

S = South, N = North

Fig. 4 Mode Shapes for Specific Frequencies

Table 1 Frequency Results from Histogram and Stabilization Diagrams

Mode Number	Mode Type (V, T)	Frequency Estimates [Hz]	
		AFDD and ERA-NExT Histogram	SSI-COV
1*	V	0.295	–
2*	?	0.350	–
3	V	0.825	0.809
4	T	0.905	0.895
5	V	1.215	1.221
6*	?	1.235	–
7*	T	1.270	–
8*	V	1.565	–
9	V	1.585	1.572
10*	?	1.660	–
11	V	1.940	1.935
12*	T	2.105	–
13	V	2.290	2.302

V = vertical mode, T = torsional mode. * = unconfirmed by SSI-COV stabilization diagram

Conclusion

This paper presents a preliminary analysis of the modal properties of the Bear Mountain Bridge. Through the application of three output-only system identification methods, AFDD, ERA-NExT, and SSI-COV, thirteen possible modal frequencies were identified, with six cross-validated by all three methods.

Analysis of the mode shapes also validates that these are real modal frequencies. For each modal frequency found in both histogram and stabilization method, corresponding mode shape results display shapes similar to what is expected at these low frequencies from previous research on bridge system identification, for example in studies done on the Golden Gate Bridge [4]. All six mode shapes are also consistent across the three output-only models used. Limitations with the number of sensors impacts the clarity of the mode shape results, but still provides enough of a picture to make determinations about the type of mode present at each analyzed modal frequency.

Estimation of modal frequencies and determination of mode types through mode shape analysis establishes baseline modal properties for the Bear Mountain Bridge. Since changes in modal properties are now easily identified through comparison with the baseline, basic structural health monitoring can now be conducted on the bridge.

Acknowledgments We would like to thank the New York State Bridge Authority and staff at the Bear Mountain Bridge for their help planning, installing, and maintaining the sensor network used in this work.

References

1. Reynders, E. "System Identification Methods for (Operational) Modal Analysis: Review and Comparison." *Arch Computat Methods Eng* 19, 51–124 (2012). https://doi.org/10.1007/s11831-012-9069-x
2. Farrar, Charles R., and Keith Worden. Structural health monitoring: a machine learning perspective. John Wiley & Sons, 2012.
3. New York State Bridge Authority, *The Bear Mountain Bridge*. https://nysba.ny.gov/bridge/bear-mountain
4. Pakzad, S. N. and Fenves, G. L., *Statistical Analysis of Vibration Modes of a Suspension Bridge Using Spatially Dense Wireless Sensor Network*. J. Struct. Eng., vol. 135, no. 7, pp. 863–872, 2009, doi: 10.1061/(asce)st.1943-541x.0000033.

5. Brinker, R., Zhang, L., Andersen, P., *Modal Identification of Output-Only Systems using Frequency Domain Decomposition*. Proceedings of the European COST F3 Conference on System Identification & Structural Health Monitoring, June 2000.
6. James III, George H., Carne, Thomas G., and Lauffer, James P., *The natural excitation technique (NExT) for modal parameter extraction from operating wind turbines*. Sandia National Labs, No. SAND-92-1666, Albuquerque, NM (United States), 1993.
7. Reynders, E., *Uncertainty Quantification in Data-driven Stochastic Subspace Identification*. Mechanical Systems and Signal Processing, 151, April 2021.
8. Caicedo, J., *Practical Guidelines for the Natural Excitation Technique (NExT) and the Eigensystem Realization Algorithm (Era) For Modal Identification Using Ambient Vibration*. Experimental Techniques, July/August 2011.
9. Al-Rumaithi, Ayad. *NExT-ERA* (https://www.mathworks.com/matlabcentral/fileexchange/72170-next-era). MATLAB Central File Exchange.
10. Cheynet, Etienne, Jakobsen, J. B., and Snæbjörnsson, J., *Damping estimation of large wind-sensitive structures*. Procedia Engineering, vol. 199, pp. 2047–2053, January 2017, doi: https://doi.org/10.1016/j.proeng.2017.09.471.
11. Cheynet, Etienne, *Operational modal analysis with automated SSI-COV algorithm*. April 2020, doi: https://doi.org/10.5281/zenodo.3774061.

Chapter 7
Optimizing NVH Performance in Electric Vehicles Through Enhanced Modal Parameter Estimation Techniques

Mahmoud Elkafafy and Marc Brughmans

Abstract When analyzing the NVH characteristics of an electric vehicle (EV) it is important to consider the effect of the battery on the dynamic characteristics of the body. When a negative impact on the body stiffness performance is noticed, body design improvements are needed to counteract the battery structural influence. A hybrid modeling environment in which a test-based battery subsystem is coupled to an FE-based body representation is ideal for this design optimization study. Frequency based substructuring techniques are typically used in such hybrid modeling environments. Due to the large mumber of interfaces DoFs between body and battery, the mumber of involved FRFs in the coupling is high ($>$ 6000 FRFs) and this poses huge challenges to the quality of the matrix inversion in the coupling operator. In such a large FRF data set the inherent observed noise and small phase inconsistencies between the FRFs can play an important role in the matrix condition of the inversion. A smoothed approximation of the FRF is needed to provide a more stable coupling condition In this paper, the MLMM modal parameter estimation method is used to achieve this smoothed approximation of the FRFs by identifying a high-quality modal model of the battery. One challenge here is that the FRFs in the different directions are different in magnitude which could lead to a poor fit quality for those FRFs that have a lower magnitude (e.g., the implane-directions' FRFs). A balancing approach will be introduced and applied to the FRFs matrix to ensure that all FRFs will equally be weighted during the modal parameter estimation process. Another challenge here is that the battery pack is coupled to the body over a large mounting area involving a large number of coupling points. This implies that a large number of FRFs are involved in the modal parameter estimation process which induces a very challenging optimization task in the MLMM method to curve fit both the global and local dynamic responses at the same time. A 2-steps workflow that delivers a full modal model of the battery, which includes all the excitation DoFs (i.e., all the coupling DoFs), starting form a limited number of FRFs matrix 's columns (i.e., a subset of the excitation DoFs) in an efficient way will be presented and validated. The proof-of-concept of the developed method is first demonstrated on a large FRF set obtained by simulation. In a second step the method has been applied on a large set of measured FRFs.

Keywords Electric vehicle · NVH · MLMM · Modal · Substucturing

Introduction

The shift of the automotive industry to Electric Vehicles (EV) brought new challenges to the development of vehicle body. The introduction of the battery pack has deeply changed the structural and dynamic behavior of the vehicle. The body design is reviewed, and structural improvements are proposed to accommodate the heavy battery structure. In this development process an accurate simulation model is needed for both body and battery stuctures. The simulation of body structures has been around for a long time, so a lot of experience is available to build an accurate model. Less experience is around for building an accurate battery structural model. When building a full FE based body-battery system model there is a danger that the quality of the simulation results is jeopardized by a lower quality FE representation of the battery. Directions for body design improvements may be biased by a wrong simulated battery behavior. Therefore, in the cases that the battery design is fixed, and hardware is available, it would be a much better option for the body designer to represent the battery with an accurate test-based modal model.

A first step in building a test-based modal model is to perform FRFs measurements on the battery at all points that interface with the body structure. These measured FRFs can directly be used in Frequency Based Substructuring (FBS)

Mahmoud Elkafafy · Marc Brughmans
Siemens Industries Software NV, Interleuvenlaan 68, 3001 Leuven, Belgium
e-mail: mahmoud.el-kafafy@siemens.com

Brandon Dilworth et al. (eds.), *Topics in Modal Analysis & Parameter Identification II, Vol. 10*, Conference Proceedings of the Society for Experimental Mechanics Series, https://doi.org/10.13052/9788743801559-7

techniques to build a full vehicle model. However, due to the large number of interfaces degrees of freedom (DoFs) between body and battery, the number of involved FRFs in the coupling is high (> 6000 FRFs) and this poses high challenges to the quality of the matrix inversion in the coupling operator. In such a large FRF data set the inherent observed noise and small phase inconsistencies between the FRFs can play an important role in the matrix condition of the inversion [1].

In this paper, we propose a method for realizing a more stable coupling condition that is based on high quality modal model that provides an accurate smoothed approximation of the measured FRFs. Several roadblocks have to be resolved to obtain such a modal model in an accurate and efficient way giving the fact that a large number of potentially magnitude-unbalanced FRFs need to be processed. The next sections will present these challenges and discuss the developed solutions from a theoretical point of view. Afterwards, the developed solutions are applied in a CAE modeling environment to validate the workflow and provide a proof of concept. Finally, the new developed techniques are demonstrated using frequency response functions (FRFs) measured on a battery structure provided by Hyundai Motor Company (HMC).

Background and Proposed Workflow

In this section, we introduce the proposed workflow for estimating high-quality modal model of a sub-component to be coupled (e.g., the battery pack) in an efficient manner. Given that the raw FRFs matrix to be processed to reach the intended modal model involves a large number of functions (78 rows × 78 columns at each frequency line) and those functions could potentially be magnitude-imbalanced, the workflow aims to address two key aspects: 1) ensuring reasonably accurate modal synthesis for all FRFs, regardless of their magnitude level, and 2) executing the modal estimation process in a time-efficient manner. The different blocks of the proposed workflow are illustrated in Fig. 1. As shown in the figure, the workflow consists of two main steps: 1) estimating a reduced-size, in terms of reference DoFs , modal model, and 2) expanding the reduced-size modal model to the full original-size FRFs matrix.

To estimate a reduced-size modal model, a selection of a few columns (e.g., $N_{i_{Red}} = 4$, with $N_{i_{Red}}$ the reduced selected number of columns) from the full-size FRFs matrix must be made. After selecting the recluced number of columns, the first step in the workflow, estimating reduced-size participation factor vectors, consists of three main sub-steps. In the first substep, a magnitude-based balancing approach is applied to the rechuced FRFs matrix to equalize the different FRFs in terms of magnitude. In the second sub-step, the modal parameters of the component under test (CUT) are estimated in the frequency band of interest by applying the well-known Polymax and MLMM (Maximum Likelihood estimation of a Modal Model) estimators [2, 3] to the balanced reduced-size FRFs matrix. Finally, in the third sub-step, the estimated mode shape vectors, participation factor vectors, and elements of the lower and upper residual terms matrices are rescaled to make them

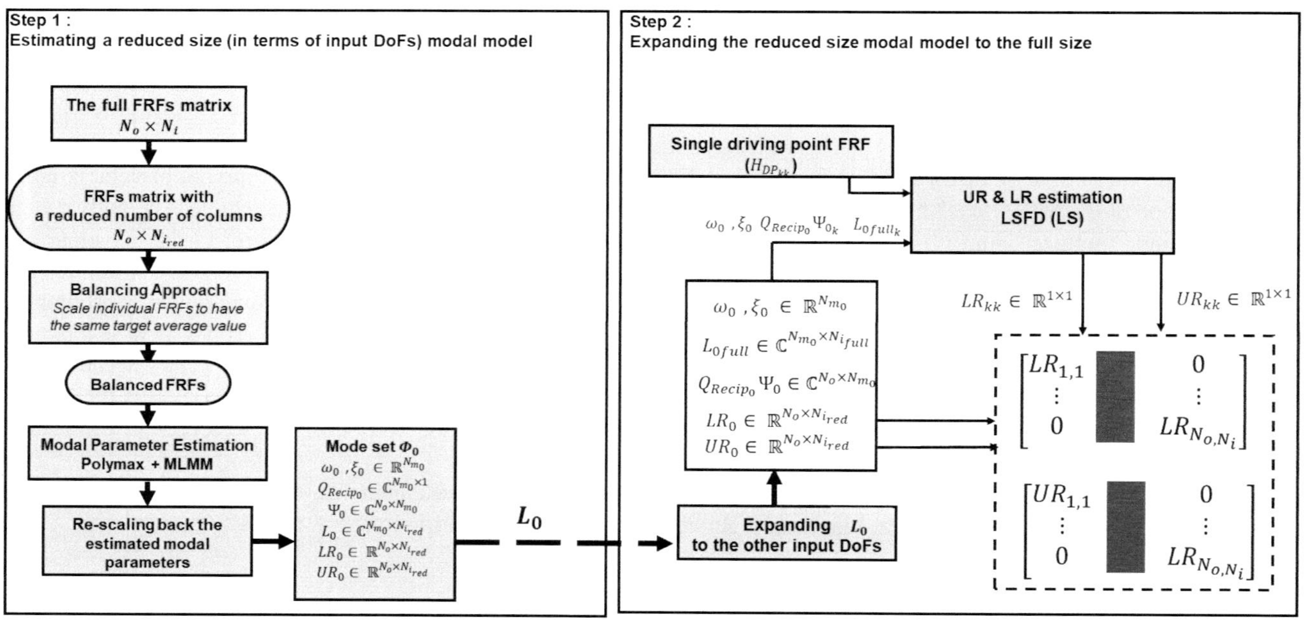

Fig. 1 The workflow of the proposed enhanced modal parameter estimation approach

consistent with the original magnitude levels of the processed FRFs. Once the reduced-size modal model is estimated, a full-size modal model is retrieved by extending the estimated participation factor vectors, the lower residual matrix, and the upper residual matrix in the second step of the proposed workflow. In the following sub-sections, futher details and background information on each sub-step are provided.

FRFs Magnitude-Based Balancing Approach

In curve-fitting techniques, such as linear least-squares (LLS)-based or nonlinear least-squares (NLS)-based fitters, there is a risk that the fitting process becomes biased toward subsets of the input data matix with relatively higher magnitudes, thereby neglecting the lower-magnitude subsets. In such a scenario, the regression model disproportionately favors certain data points, often those with higher values or greater variance, while underweighting or 'forgetting' data points with lower values. Since these are essentially least-squares-based techniques, advanced frequency-domain modal parameter estimators, such as Polymax, MLMM, and others [2–4], could face this issue if the primary imput data matrix (e.g., an FRFs matrix) is imbalanced in terms of the magnitude levels of its individual entries. This imbalance in the magnitude levels of the measured response functions could be due to several factors: differences in the stiffness of the component under test (CUT) across vanious directions or paths, the inclusion of both translational and rotational responses in the measured DoFs (e.g., translational and rotational accelerations), or the involvement of quantities with mixed units in the measured FRFs matrix (e.g., vibroacoustic modal test where structural and acoustic FRFs are acquired).

In the context of modal model-based structural coupling (or component mode substructuring CMS), it is crucial to ensure that all the FRFs at the coupling DoFs, specifically those involving co-located measurements of responses and forces, are accurately represented by the identified modal model. In an electric vehicle battery-like structure, the response magnitudes in the x and y (i.e, longitudinal and lateral) directions are relatively lower than those in the z (i.e., vertical) direction due to the battery's geometry. This results in lower-magnitude FRFs in the x and y directions and higher-magnitude FRFs in the z-direction. Figure 2 shows typical driving point FRFs extracted from an EV battery's finite element model (FEM), where the difference in magnitude across the x, y, and z directions is clearly visible. Performing a modal parameters estimation using an FRFs matrix that contains such different in magnitude functions could end with biased results, as not all FRFs in the input matrix would be appropriately considered. The FRFs in x and y-directions may be underweighted during the modal fitting process compared to those in the z-direction, leading to enroneous modal synthesis for those FRFs. This would certainly result in inaccurate body-to-battery coupling result. A similar issue of unbalanced FRFs can occur on the car body side. For example, a DoF measured on the roof of the car body, which is an important DoF for visualizing global modes, will have a significantly higher amplitude than a DoF measured on the driver's seat rail or a DoF at the body-to-battery connection points, which are used for target settings and structural coupling, respectively. This imbalance could mislead the modal estimation technique, causing it to focus on the higher-magnitude DoFs while neglecting the ones that are more critical for analysis (i.e., the coupling and target settings DoFs).

In this work, a magnitude-based balancing approach will be implemented to ensure that all individual FRFs will appropriately be considered by the Polymax and MLMM modal parameter estimators during the modal parameter estimation step. The implemented balancing approach aims to scale all individual FRFs so that they have the same target average value (e.g., mean, median, RMS value) equals to 1 over the selected fiequency band of interest. This is achieved by scaling each function in the FRFs matrix with its calculated 'selected' central tendency criterion Γ_{ot} with $o = 1, 2, \ldots, N_o$ and $i = 1, 2, \ldots, N_t$, where N_o and N_l are the mumber of outputs (responses DoFs) and inputs (excitation DoFs) respectively. The effect of this scaling mechanism is shown in Fig. 3 where the unbalanced FRFs (cf. Fig. 2) are now getting closer to each other in terms of magnitude level. Scaling the magnitude of the raw FRFs will consequently influence the magnitudes of the estimated mode

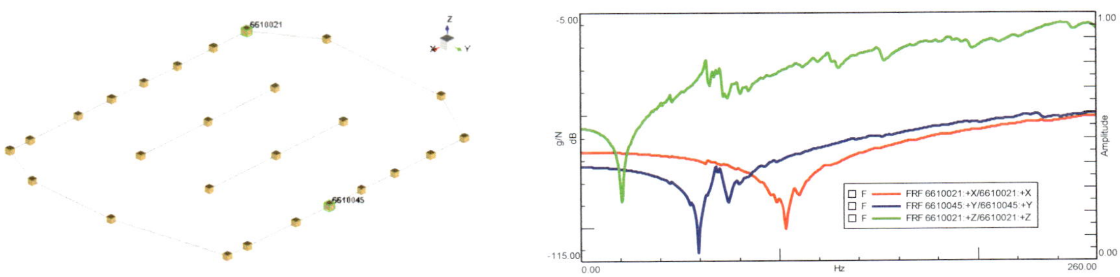

Fig. 2 Right: typical driving point FRFs in different directions extracted from an EV battery's FEM. Left: the geometry of the battery of an EV highlighting the selected driving point locations

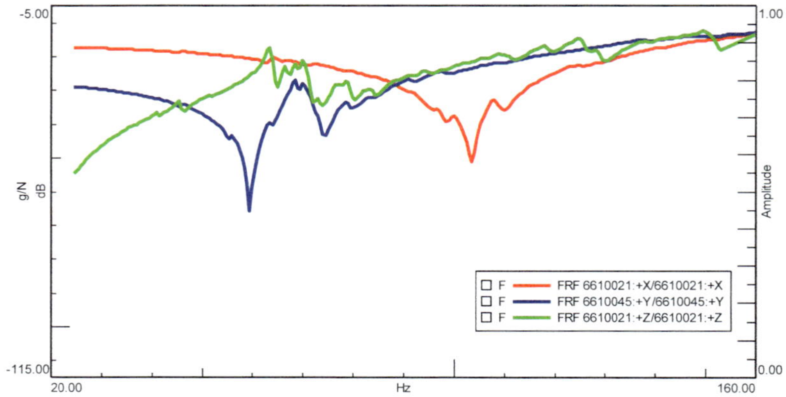

Fig. 3 The effect of the balancing approach on the magnitude level of originally unbalanced FRFs (cf. Fig. 2)

shapes, participation factor vectors, and the lower and upper residual terms (cf, equation (1)). This means that the estimated modal vectors (i.e, mode shapes and participation factors vectors), along with the residual terms, will need to be rescaled back to their expected correct magnitude levels that are consistent with the magnitude levels of the original raw FRFs.

To achieve this, the scaling coefficient Γ_{oi} used to scale an individual FRF H_{oi} needs to be split into two coefficients $\Gamma_{ol} = \alpha_o \beta_i$, with $o = 1, 2, \ldots, N_o$ and $i = 1, 2, \ldots, N_i$. The α_o's coefficients will be used to rescale the mode shape elements, while the β_i's coefficient will be used to rescale the participation factor vectors for each identified mode. The Γ_{oi} coefficients will be used to rescale the entries of the lower and upper residual matrices. Expressing the relationship $\Gamma_{oi} = \alpha_o \beta_l$ in a set of linear equations, one can determine the α_0 and the β_1 coefficients by solving a linear least-squares problem considering the constraint for preserving the FRFs reciprocity property if needed.

Modal Parameter Estimation: Polymax & MLMM Estimators

In this step, the Simcenter MLMM modal parameter estimator [3] will be applied to the balanced FRFs matrix to estimate a modal model that can represent the measured FRFs with a reasonable accuracy within a specified frequency band. To expedite the modal estimation step, the FRFs matnix with the full set of columns (in our case this corresponds to $N_i = 78$) will not be used Instead, a significantly reduced FRFs matrix with fewer columns (e.g., $N_{i_{Red}} = 4$) will be provided as input data to the MLMM estimator. There is no specific theory to guide users in selecting the rechuced number of references DoFs (in terms of quantity and location). However, a rule of thumb from domain knowledge suggests that the selected reference DoFs should ensure excitation in all directions (vertical, lateral, and longitudinal) and should not be concentrated on one side of the component under test (CUT).

The modal model given by Eq. (1) decomposes the frequency response functions (FRFs) matrix of a dynamical system $\boldsymbol{H}(s_k) \in \mathbb{C}^{N_0 \times N_i \times N_f}$, with $s_k = j\omega_k$ the complex fiequency at frequency line k where $k = 1, 2, \ldots, N_f$, as a linear sum of the different N_m vibroacoustic modes represented by their corresponding modal parameters $\psi_r \in \mathbb{C}_0^{N_o \times 1}$ the rth mode shape column vector, $\lambda_r \in \mathbb{C}$ the rth pole from which the natural fiequency and the damping ratio are deduced, and $L_r \in \mathbb{C}^{1 \times N_t}$ the rth participation factor row vector.

$$\boldsymbol{H}(s_k) = \left(\sum_{r=1}^{N_m} \left(\frac{\psi_r L_r}{s_k - \lambda_r} + \frac{\psi_r^* L_r^*}{s_k - \lambda_r^*} \right) \right) + \frac{\boldsymbol{LR}}{s_k^2} + \boldsymbol{UR} \tag{1}$$

In Eq. (1), (.)* stands for the complex conjugate of a complex-valued parameter, $\boldsymbol{LR} \in \mathbb{R}^{N_0 \times N_l}$ and $\boldsymbol{UR} \in \mathbb{R}^{N_0 \times N_l}$ are the lower and upper residual terms respectively that are used to model the influence of the modes that are located respectively below and above the fiequency range of interest considering a displacement over force FRF format. The upper residual term (***UR***), also called the stiffness residual, is essentially a constant that models the asymptotic contribution of modes located at frequencies far above the frequency band of interest. This term, particularly its diagonal elements, is crucial for accurately modeling the dynamics of a coupling DoF (i.e, a driving point DoF), as it compensates for the local stiffless contribution that is not captured by the global modes within the frequency band of interest. The lower residual, also known as the mass residual, models the asymptotic contribution of modes located far below the fiequency band of interest. This contribution decays with the square of the fiequency.

The MLMM modal parameter estimation method [3] is a multivariable (i.e., MIMO) frequency-domain method that uses the modal model (1) as a parametric model to represent the measured FRFs over a chosen frequency band. The MLMM method differs from the existing modal parameter estimation methods in the sense that it directly optimizes the modal model, represented by equation (1), rather than optimizing a high order rational fiaction polynomial-based model and reducing it in a subsequent step to a modal model. Such model reduction would most likely lead to a degradation in the quality of the fit between the measured data and the final estimated modal model In the MLMM method, the Levenberg-Marquardt optimization algorithm tunes the parameters of the modal model (1) to minimize a quadratic-like cost function (i.e., sum of the squared enror between the true data and the data synthesized by the modal model) so that a best match between the modal model (1) and the measurements is obtained. In the optimization process, the user has also the option to apply some physically motivated constraints to the modal model, e.g. FRFs reciprocity constraint, real mode shapes constraint, and the negative mass sensitivity (NMS) characteristic constraint. The inputs needed from the user in the MLMM step are the maximum number of iterations for the iterative Levenberg-Marquardt optimization process, the choice of including/excluding the lower and upper residual terms, and if constraints are to be applied to the modal model. In this work, the MLMM estimator will be applied to estimate a constained modal model, where FRFs reciprocity, real mode shapes, and NMS constraints are imposed on the identified modal model. The FRFs reciprocity constraint is crucial for the subsequent expansion step, where the modal model will be expanded to match the full size of the original FRFs matnix (i.e., from $N_{i_{Red}} = 4$ to $N_i = 78$). The NMS and real mode shape constraints are imposed not only to preserve the physicalness of the identified modes and facilitate mode shape visualization but also to ensure that the identified mode set is exportable to Simcenter Nastran, where the battery-to-body coupling activities will be carried out.

Since MLMM runs a nonlinear optimization process, it requires initial values for the system poles $\lambda_r \in \mathbb{C}$ and the participation factors $L_r \in \mathbb{C}^{1\times N_l}$ for each identified mode within the flequency band of interest. The well-known linear least-squares-based modal estimator Polymax [2] will be used to generator these initial values for the MLMM estimator. In a nutshell, the Polymax estimator requires FRFs as primary data and identifies, by fitting those FRFs, the numerator and denominator matrix polynomial coefficients of a right matrix fraction description (RMFD) polynomial transfer function model defined with a model order p set for both the numerator and the denominator polynomials. Those coefficients are identified by solving a linear least squares problem with a cost function given as the squares of the (lineanized) fit enror between the measured FRFs and the RMFD model. Once the denominator matrix polynomial coefficients are estimated, the eigenvalues λ_r and the modal participation factors L_r are calculated by computing the eigenvalue decomposition of their companion matrix. An important user choice is the polynomial model order p, which is related to $n = 2N_m$, the number of poles in the system as: $n = pN_i$. Rather than fixing the model order a priori, a whole range of models with increasing model orders (up to a maximum model order p) is calculated. The poles conesponding to a certain model order are compared to the poles of a one-order-lower model If their differences in terms of frequency, damping, and eigenvectors are within pre-set limits, the pole is labelled as a stable one. The spurious numerical poles will not stabilize at all during this process and can be sorted out of the set of poles more easily, e.g., by interpreting so-called stabilization diagrams. A stabilization diagram is a 2D plot showing all poles from all identified models, sorted according to frequency values (x-axis) and model order (y-axis), cf. Figure 16 for instance. In such diagram, the stable poles will also appear as clusters and a representative of each cluster can be selected for the final mode set. One big advantage of the Polymax estimator is that it delivers a very clear stabilization diagram which helps the user to easily select the physical modes and to set up the proper model size of the analyzed system. Altematively, the constructed stabilization diagram can be automatically interpreted instead [5] where the mumber of stable pole clusters determines the model order and a pole estimate (and the corresponding participation factors) for each cluster is calculated. The inputs needed from the user in the Polymax step are the maximum polynomial order p and the fiequency band of interest $[f_{\min}, f_{\max}]$.

As mentioned earlier, the balanced FRFs-based estimated modal parameters (mode shape vectors, participation factor vectors, lower residuals, and upper residuals) need to be rescaled to make them consistent with the magnitude level of the original imbalanced FRFs. The estimated scaling coefficients α_o's and the β_l's (cf. the FRFs balancing section) will be used to do this rescaling process.

Modal Model Expansion

After estimating a reduced-size (in terms of the number of reference DoFs) reciprocal modal model using the Polymax and MLMM modal estimators, the estimated participation factor vectors, along with the lower and upper residual matrices, will be expanded to the fill number of reference DoFs (i.e., from $N_{i_{Red}} = 4$ to $N_i = 78$). A reciprocal modal model means that the lower and upper residual matrices are symmetric, and for each identified mode, the mode shape coefficient at an imput DoF is proportional to the participation factor at that imput DoF, up to a scaling factor (i.e, the reciprocal scaling factor). When the reciprocity constraint is applied during the MLMM iterations, the MLMM estimator produces the following modal model:

$$\boldsymbol{H}_{Red}(s_k) = \left(\sum_{r=1}^{N_m} \left(\frac{Q_r \phi_r L_{Red_r}}{s_k - \lambda_r} + \frac{Q_r^* \phi_r^* L_{Red_r^*}^*}{s_k - \lambda_r^*} \right) \right) + \frac{\boldsymbol{LR}_{Rec}}{s_k^2} + \boldsymbol{UR}_{Rec} \tag{2}$$

with, for the rth mode, $Q_T \in \mathbb{C}$ the reciprocal scaling factor, $\phi_r \in \mathbb{C}^{N_o \times 1}$ the mode shapes vector in which the shape coefficient conresponds to an input DoF considered to be proportional to the participation factor at that input DoF up to a scale factor Q_r, $L_{Red_i} \in \mathbb{C}^{1 \times N_{i_{Red}}}$ the reduced size participation factors vector, $\boldsymbol{LR}_{Rec} \in S^{N_0 \times N_{i_{Red}}}$ and $\boldsymbol{UR}_{\text{Rec}} \in S^{N_0 \times N_{i_{Red}}}$ the reduced size symmetric lower and upper residual matrices. In case the real shape and the negative mass sensitivity constraints are also imposed, the product $Q_r \phi_r L_{Red_r}$ becomes an imaginary-valued matrix, where the element conresponding to a driving point DoF is imposed to be a negative imaginary value. Having identified the reciprocal modal model (2) with the reciprocal scaling factors Q_r, the expanded participation factors vector $L_r \in \mathbb{C}^{1 \times N_i}$ for each identified mode r is simply the transpose of the corresponding mode shapes vector:

$$L_r = \phi_r^T \in \mathbb{C}^{1 \times N_0} \tag{3}$$

For the lower and upper residual terms (***LR*** & ***UR***), it is not feasible to fully expand those matrices to obtain all $N_o \times N_o$ elements, as was done for the modal kernel term (also known as the global flexable modes term). However, neglecting these terms can lead to significant errors in modal synthesis, especially if lower and/or higher-fiequency modes are truncated. This issue is particularly critical for a driving point degree of freedom (DoF). The global flexible modes and the higher-fiequency truncated modes, represented by the upper residual term (***UR***), both contribute to the local flexability at a driving point degree of freedom (DoF). The more significant the higher-fiequency truncated modes are, the more critical the ***UR*** contribution becomes to local flexability. Consequently, the diagonal elements of the upper residual term (***UR***) are essential for achieving an accurate modal synthesis of a frequency response function (FRF) of a driving point. Higher-frequency truncated modes contribute more significantly to local flexability, i.e., the diagonal elements of the upper residual term (***UR***), rather than to global flexibility, which is primanily captured by the contributions of global flexible modes. Therefore, a reasonable approximative solution for expanding the ***UR*** matrix is to focus only on the diagonal elements. This approach would provide an accurate modeling of local stiffness for the driving point degrees of freedom (DoFs), which are essentially the coupling DoFs between the car body and the battery structure. For the lower residual (***LR***) matrix, the same scheme used for the upper residual (***UR***) matrix will be followed, focusing on extending only the diagonal elements. The contribution of the ***LR*** matrix to the overall dynamics of the component under test is found to be minor compared to that of the ***UR*** matrix. This is because the analyzed frequency band typically includes the first global mode, indicating that no fundamental modes exist below the frequency band of interest. The full diagonal elements of the upper residual matrix ***UR*** and the lower residual matrix ***LR*** are determined by solving a linear least-squares problem described by a reciprocal SISO modal model based on Eq. (2) and incorporating the reciprocity property expressed in Eq. (3). In this set of equations, involving only co-located DoFs %, the only unknows are the corresponding lower and upper residuals: LR_{oo} and UR_{oo}.

This estimated and expanded modal model is designed to capture both global and local flexability. It incorporates not only the contributions from the global fundamental flexible modes (the modal kernel term) but also those from the higher-fiequency truncated modes. These higher-fiequency modes play a crucial role in conecting local stiffness for all measured coupling degrees of freedom (DoFs).

Analysis & Results: Simulation Data

In this section, the workflow mentioned above will be validated using a dataset consisting of Frequency Response Functions (FRFs) extracted from a finite element model (FEM) of an EV battery, which contains 927712 nodes (see Fig. 4, left). The FRFs were simulated at 26 body-to-battery connection points, each with 3 tanslational degrees of freedom (DoFs) in the x, y, and z directions, resulting in a square FRF matrix with 78 rows (outputs) and 78 columns (inputs) at each frequency line. Figure 4, right, illustrates some typical FRFs simulated by the battery FEM. In our analysis, the frequency range was limited to $25 - 160$ Hz, as the FEM had been experimentally calibrated only up to this fiequency at the time of writing this paper. According to the FEM, the battery structure is expected to exhibit approximately 114 flexable modes within the $25 - 160$ Hz frequency range. Starting from the $78\ (N_o) \times 78\ (N_l)$ FRFs matrix extracted from the battery model, the goal is to identify a modal model that captures both the global flexible modes and the local flexability at the connection DoFs in this frequency band. The identification of that modal model must ensure both accuracy and relatively fast computation. The challenges arise from two main factors: 1) the number of flexable modes should be minimized while maintaining a good fit, and 2) local flexibility must be included for all the coupling degrees of freedom (DoFs) and this requires including all 78 columns of the FRFs matrix in the modal fitting process, which is computationally impractical.

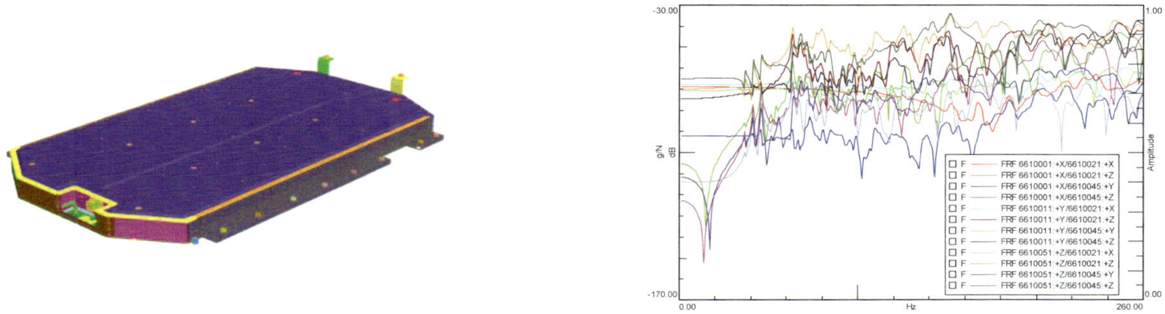

Fig. 4 The battery pack FEM (left) and some typical FRFs at connection points (right)

As mentioned in the workflow description section, the first step is to select a few columns from the full-size FRFs matrix, which originally contains 78 columns. By reducing the number of columns, the goal is to significantly speed up the modal estimation process of the global flexible modes. Four columns are selected, two conesponding to each of the two excitation locations shown in Fig 2-left (two columns per location). This results in a 78×4 FRFs matrix at each frequency line, which is then processed through the implemented FRFs balancing technique to equalize the magnitude levels of the 312 FRFs considered.

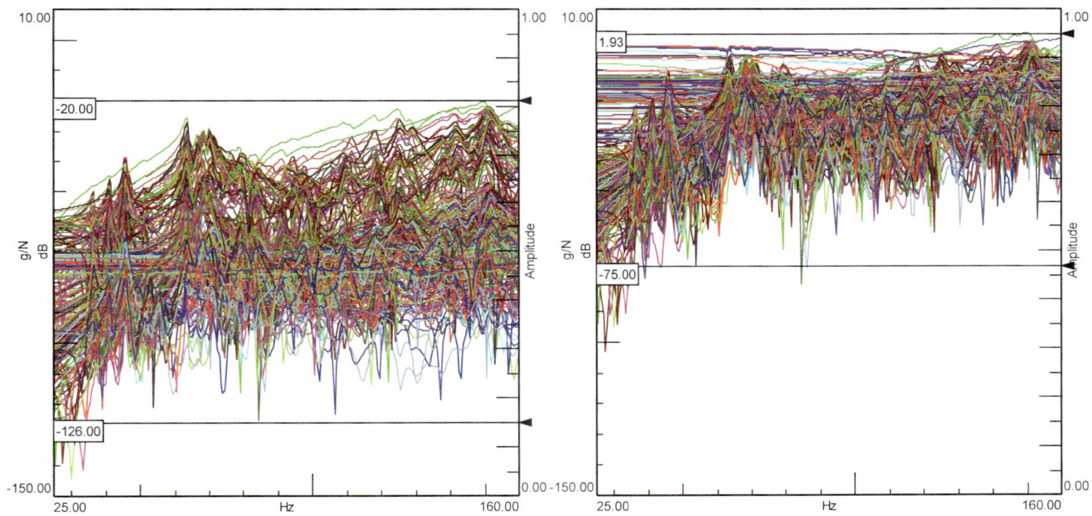

Fig. 5 Unbalanced FRFs matrix (left) & Balanced FRFs matrix (right)

Figure 5 shows the FRFs before and after applying the balancing technique. The balanced FRFs are visibly closer to each other in terms of their magnitude level compared to the unbalanced ones. This is also objectively reflected in the reduced dynamic range of the FRFs matrix after balancing, as shown in the figure.

Having balanced the FRFs matrix, the MLMM modal parameter estimator is then applied to the balanced matrix to optimize the modal model represented by Eq. (1) imposing the FRFs reciprocity, real mode shapes, and the negative mass sensitivity (NMS) constaints on the identified model. The MLMM optimization process is set to rum for 20 iterations, as convergence in the cost function evolution, as shown in Fig. 7, is observed starting from iteration 15. To start the iterations of the nonlinear optimization process in the MLMM estimator, reasonable initial values for the pole and the participation factors vector (i.e., λ_r, L_r parameters in Eq. (1) for each identified mode are needed. To estimate these initial values, the Polymax estimator, with a model size of 64, is applied to the balanced FRFs matrix. The Polymax-constructed stabilization chart is shown in Fig. 6, from which 48 modes are selected using the automatic modal parameter selection option [5]. The MLMM estimator then optimizes the initial values of the modal parameters of those 48 modes by minimizing the enror between the modal model (1) and the true (or the reference) FRFs. To evaluate the effectiveness of the FRFs balancing technique, the modal parameter estimation process, consisting of Polymax followed by 20 ML MM iterations, was conducted twice: first using the unbalanced (raw) FRFs matrix, and then using the balanced FRFs matrix.

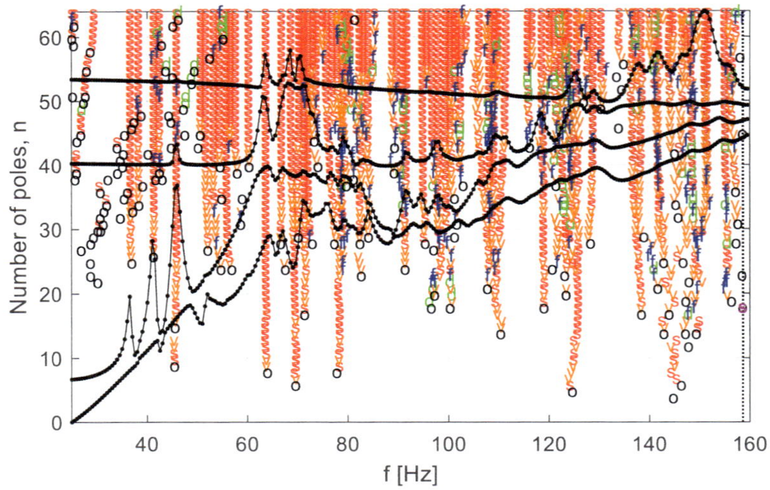

Fig. 6 Polymax stabilization chart (the black curves represent the 4 Complex Mode Indicator Function CMIIFs)

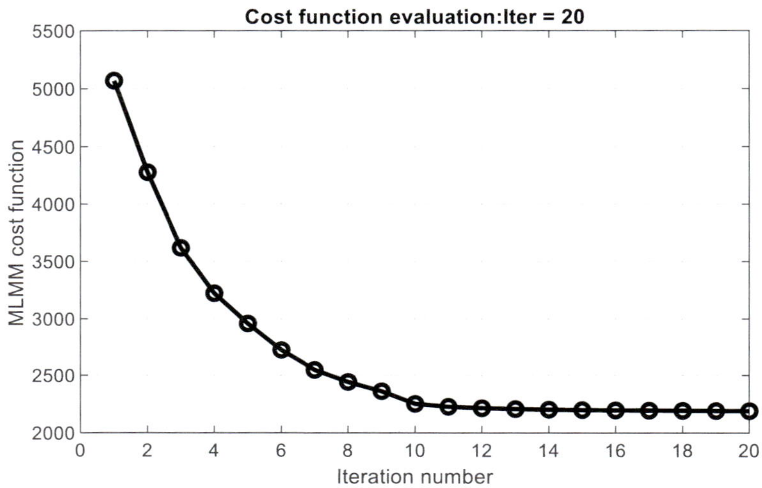

Fig. 7 MLMM iterations (the enror between the modal model and the true FRFs is decreasing over the different MLMM iterations)

Figure 8 shows color maps representing the relative modal synthesis enror percentage for each FRF, comparing two cases: with and without balancing the raw FRFs matrix before doing the modal estimation step. The relative modal synthesis enor is calculated as the sum of the errors between the reference FRF and the synthesized FRF across all frequency lines, divided by the reference FRF. As seen in these color maps in Fig. 8, balancing the FRFs matrix before performing modal parameter estimation improves the quality of the resulting modal model. This is evident from the lower relative modal synthesis enror percentage when using a balanced FRFs matnix compared to an imbalanced one. The improvement is particularly noticeable for FRFs with lower magnitude levels, such as those conesponding to excitation in the x and y directions. The modal synthesis of selected FRFs is presented in Figs. 9 and 10. Figure 9 illustates the modal synthesis for FRFs that originally have a low magnitude, while Fig. 10 shows the synthesis for FRFs with a higher magnitude. The results clearly demonstrate that balancing the FRFs matrix before performing modal parameter estimation improves the synthesis for lowmagnitude FRFs, without negatively affecting the synthesis for high-magnitude FRFs. Figure 11 displays the mode shapes of some extracted global flexible modes.

Once the reduced-size (78×4) reciprocal modal model is identified, the paticipation factor vectors, along with the diagonals of the lower and upper residual matrices, are extended using the expansion approach described earlier in the paper. This results in the final modal model which matches the size of the original full-size 78×78 FRFs matrix. Figure 12 judges

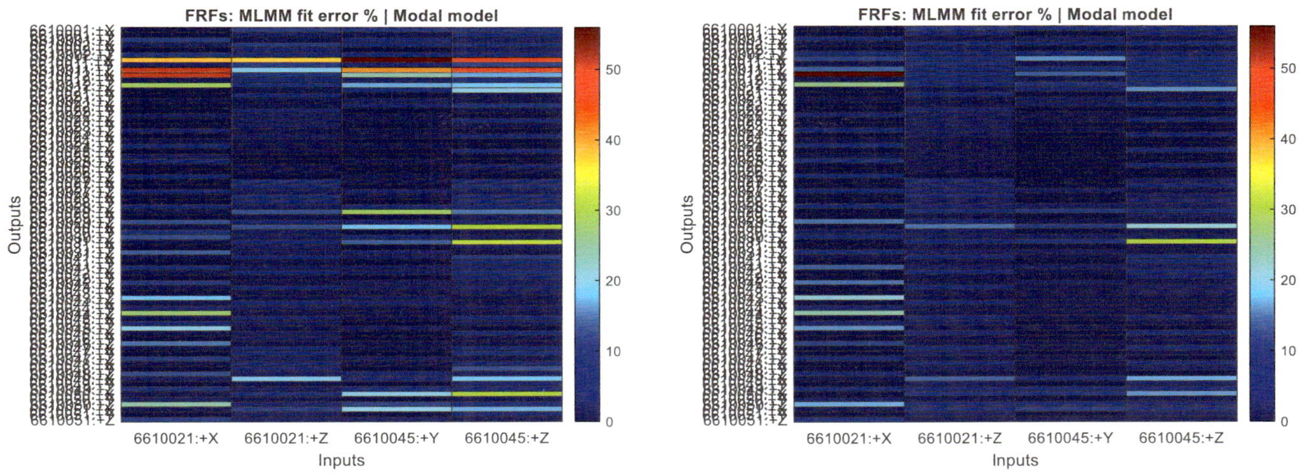

Fig. 8 The modal synthesis error % between the synthesized FRFs and the true ones without balancing the FRFs matrix (left) and with balancing the FRFs matrix (right)

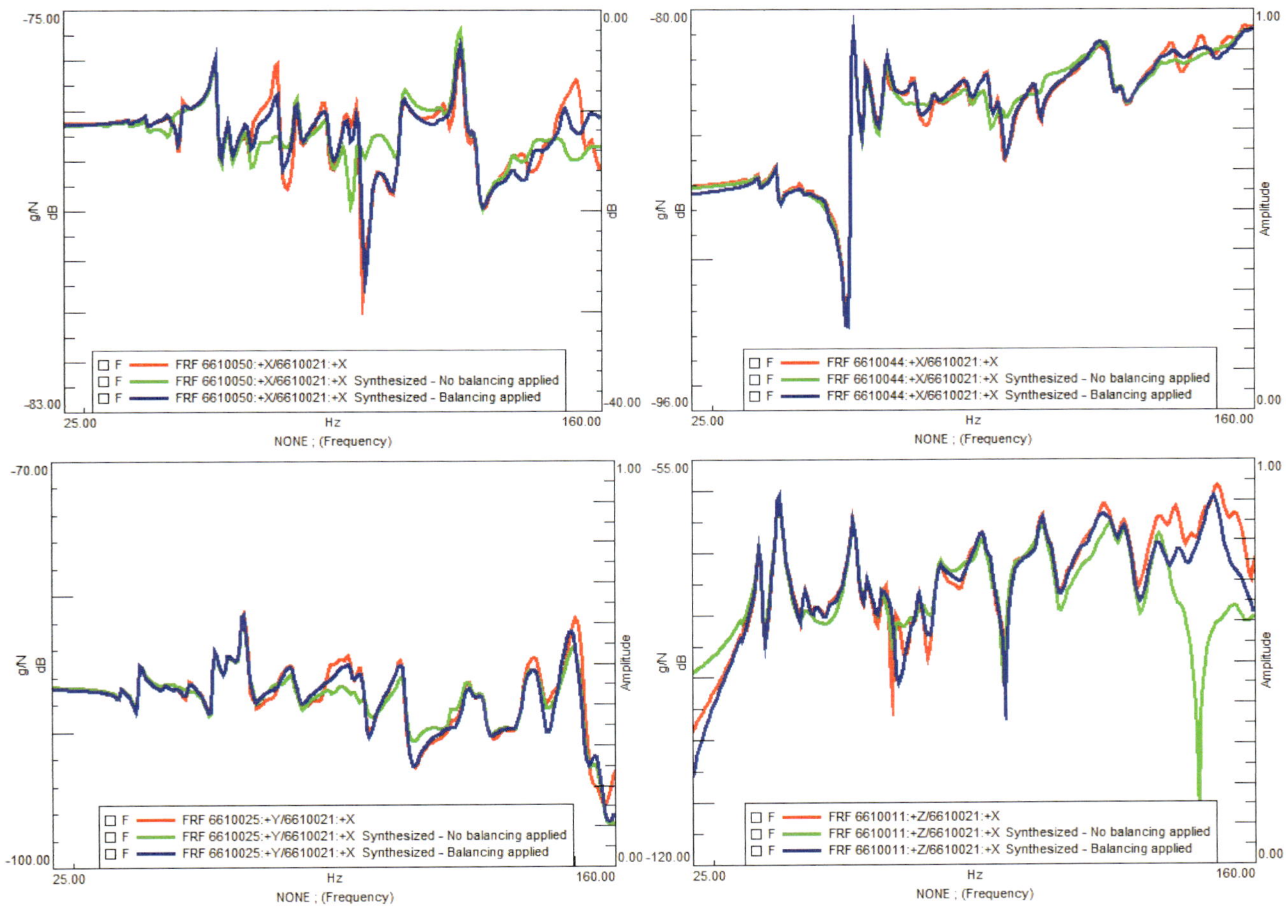

Fig. 9 The effect of the balancing approach on the quality of the final modal synthesis for some FRFs that originally have a low magnitude level

the accuracy of the expanded modal model by examining the modal synthesis of the FRFs of one driving point (in x, y, and z-directions) that were not initially included in the reduced-size FRFs matrix used to perform the modal parameter estimation. The figure illustrates the modal synthesis using three different approaches: 1) Modal estimation is performed directly with

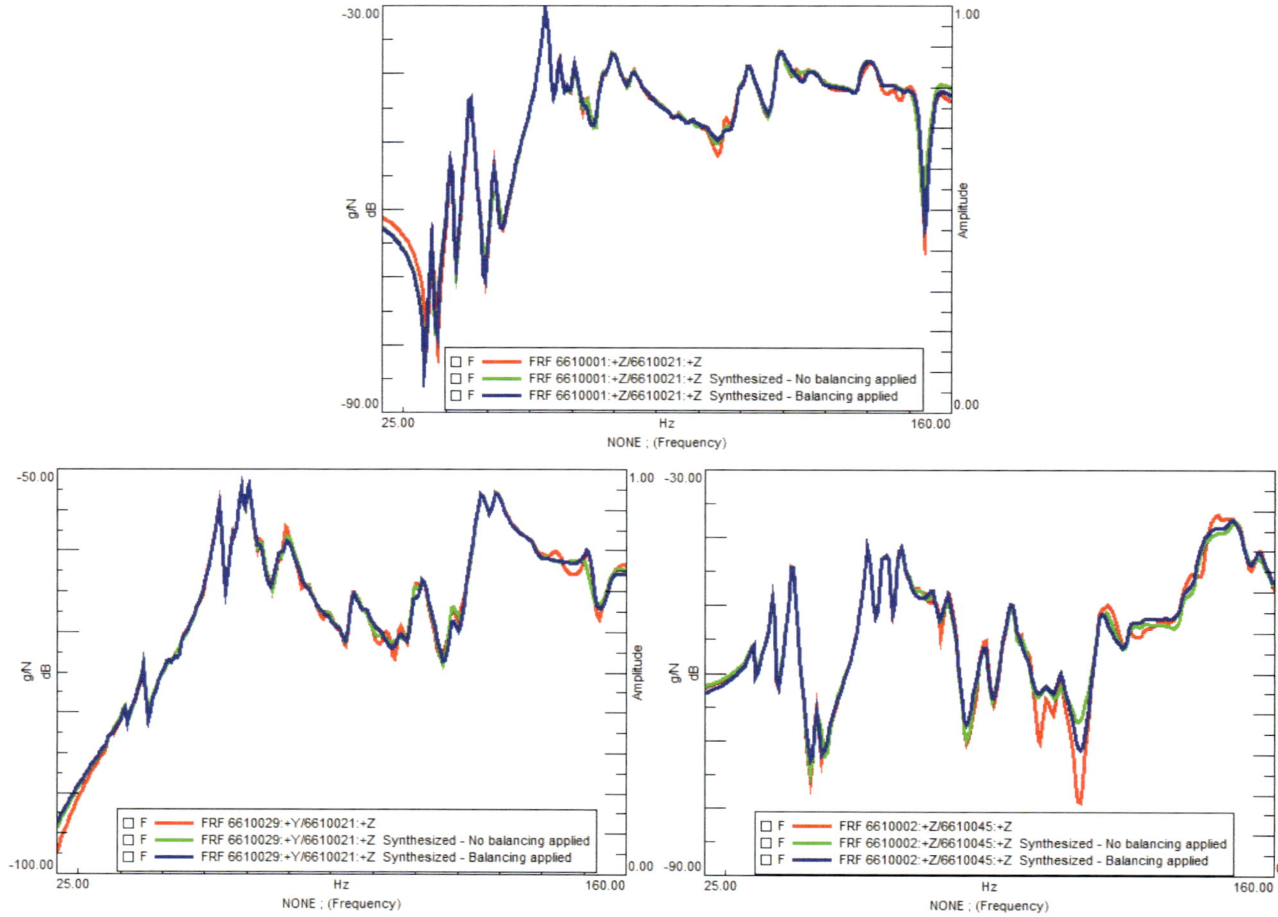

Fig. 10 The effect of the balancing approach on the quality of the final modal synthesis for some FRFs that originally have a high magnitude level

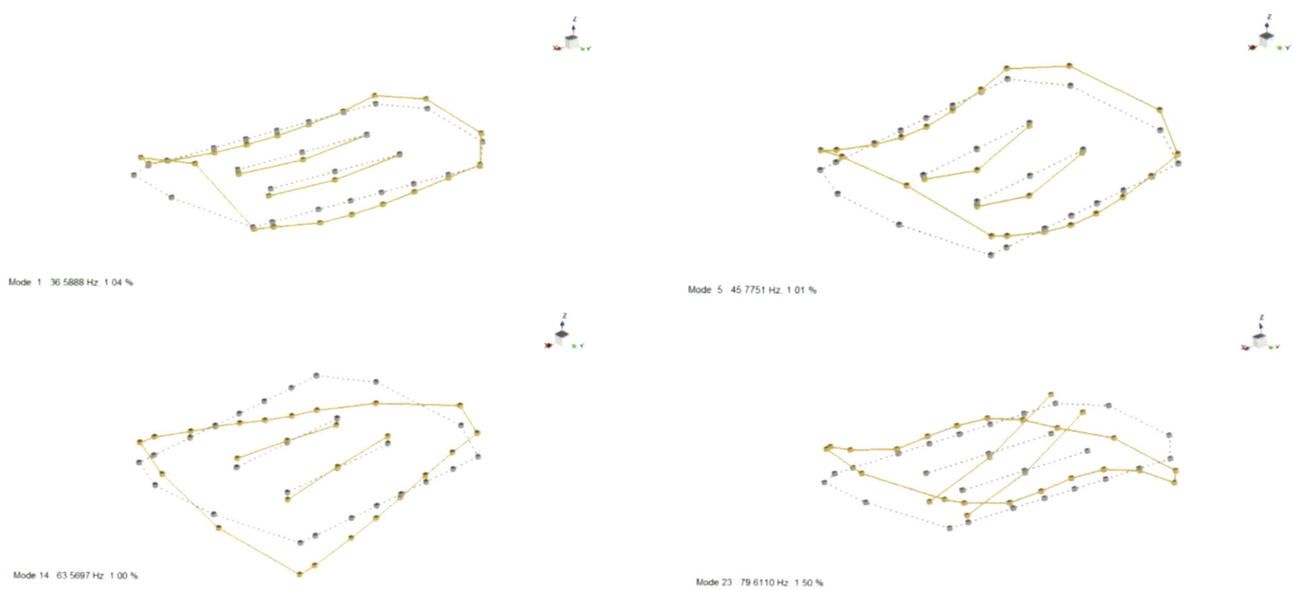

Fig. 11 Examples of estimated global flexible mode shapes

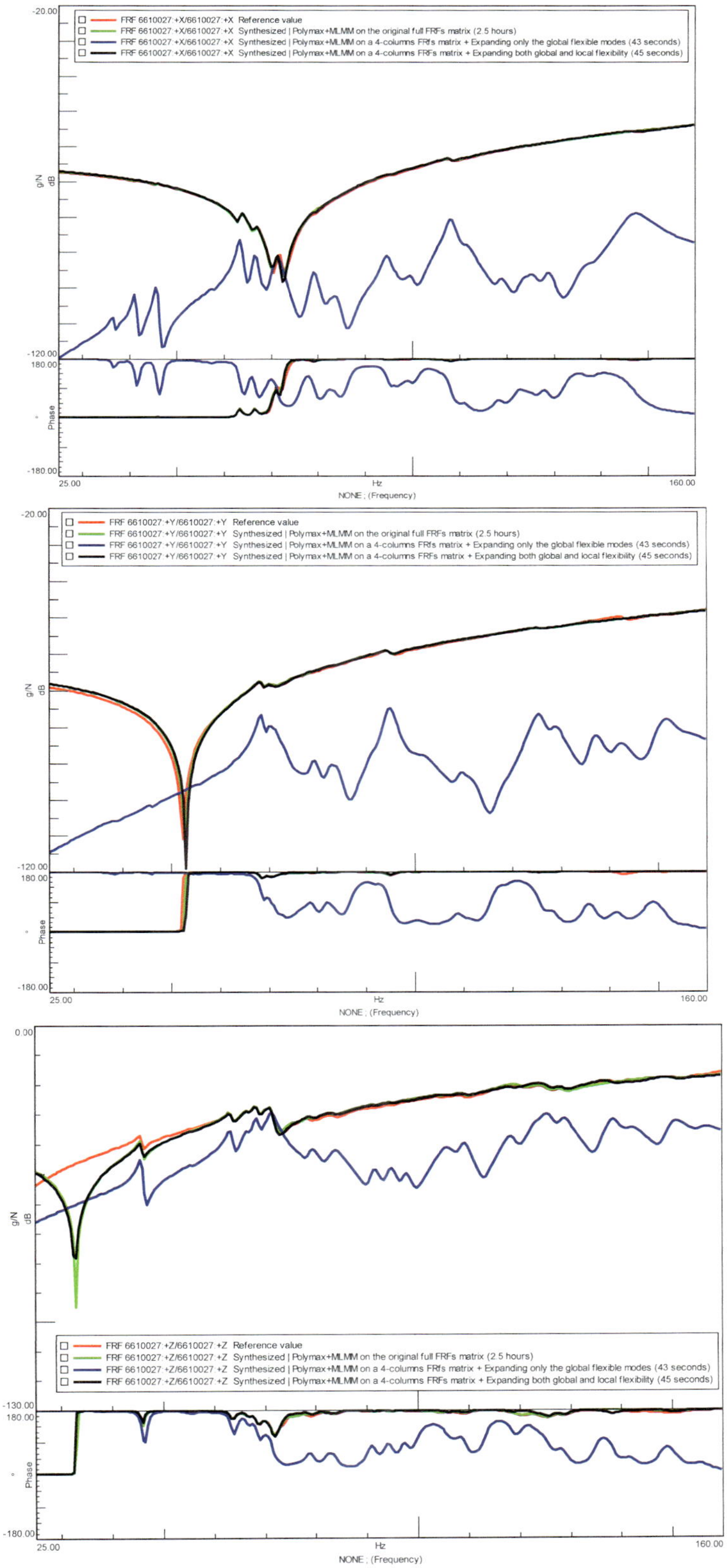

Fig. 12 The modal synthesis of FRFs of a connection point (in x, y, and z-directions) obtained from 3 different approaches for each the computational time is indicated in the figure legend (the shown FRFs were initially not included in the rechuced-size FRFs matrix that was used to perform the modal parameter estimation of the global flexable modes)

the full-size FRFs matrix (78×78), requiring no expansion; 2) Modal estimation is conducted with a reducedsize FRFs matrix (78×4), where only the contributions of global flexible modes are expanded while ignoring local flexibilities (i.e., the diagonal elements of the UR matrix); and 3) Modal estimation is carried out with the same reduced-size FRFs matrix (78×4), expanding contributions from both global and local flexibilities. The computational time for each approach is provided in the figure legend. The results from approach 2 in the Fig. 12 clearly demonstrate the importance of including local flexabilities. Ignoring this contribution leads to a significant degradation in the modal synthesis of the final identified modal model. Approach 1, which uses the full FRFs matrix, produces a modal model that closely fits the true FRFs. However, the computational time required for this approach is substantial (2.5 hours), making it impractical, especially for applications in vehicle design development. The lengthy computation time for approach 1 arises from the large mumber of parameters (27240) that must be optimized simultaneously during the MLMM iterations.

Figure 12 shows that approach 3 outperforms both approach 1 and approach 2 in terms of computational time and accuracy of the modal synthesis. Approach 3 delivers an accurate full-size modal model that accounts for both global and local flexibilities in just 45 seconds, significantly outperforming the other two approaches. This efficiency is evident from the quality of the modal synthesis achieved in a relatively short computational time. The reduced computational time of approach 3 is due to the optimization of only 8592 parameters, a significantly smaller number compared to the 27240 parameters in approach 1 , resulting in a more efficient process.

As a final validation step in this section, Fig. 13 illustrates the body-to-battery coupling results for a selected target DoF FRFs when the MLMM-based identified battery modal model, considering only 48 global flexible modes, is coupled to the vehicle body modal model (referred to as Assembly 2 in Fig. 13). These results are compared to the reference case modal system assembly (Assembly 1 in Fig. 13), which utilizes the full mode set identified by FEM for both the body and battery within the frequency band of interest. This full mode set includes 1,032 global flexible modes for the vehicle body and 114 for the battery, along with rigid body modes and static compensation modes to account for the truncated higher-frequency modes, as shown in Fig. 13. It is worth to mention that the Assembly 1 results were compared against a full body FEbattery FE assembly and the results were identical. Two key conclusions can be drawn from the results. First, the figure clearly demonstrates that the battery significantly impacts the NVH characteristics of the vehicle body. The system represented by the green curve (BIP - no battery coupled) exhibits a notably different response compared to the system represented by the red curve (battery coupled to the body with the full mode set). This highlights the importance of considering body-to-battery coupling when assessing the NVH performance of an electric vehicle (EV). Second, the coupling result obtained using the MLMM-based reduced-order modal model, considering both global and local flexibilities, albeit with only 48 flexible modes, closely matches the result from the full system assembly (Assembly 1, which includes 114 flexible modes for the battery). These findings validate that the proposed workflow effectively delivers a reduced-order modal model capable of adequately capturing the contributions of both global flexible modes and local flexibilities at the coupling degrees of freedom (DoFs).

Analysis & Results: Experdmental Data

In this section, the developed workflow will be tested and validated using test data that were measured churing a modal test of the battery structure. In this modal test, the battery structure was free-fiee suspended in bungees (cf. Fig. 14) and excited up to 2 kHz at 78 coupling DoFs (26 body connection points in x, y and z-directions) using the Simcenter Qsources hardware integral shakers (Qish-Integral shaker) mounted on Qtax excitation adaptor. Frequency response functions were measured consecutively under different multiple angles as shown in Fig. 14. The angled FRF set is processed to reconstruct the FRFs with the orthogonal imputs of the global axis system. The acceleration responses were simultaneously acquired at the 78 coupling DoFs using modal accelerometers PCB 356A15. This results in a $78\,(N_0) \times 78\,(N_0)$ FRFs matrix at each frequency line in the excited frequency range. Some typical measured FRFs are shown in Fig. 14.

In our analysis, we have targeted our fiequency application range $[100 - 1000$ Hz$]$ to start from the validated FB battery upper frequency limit (< 160 Hz) and go up to the mid frequency range (1 kHz) where CAE body models can still be represented by FEM. A reduced FRF matrix consisting of only four columns is selected from the full FRF matrix to perform the modal estimation of the global flexible modes within the frequency band of interest more efficiently. Before applying the Polymax and MLMM modal parameter estimators to the reduced FRF matrix, the matrix is first balanced using the implemented balancing technique. The effect of the balancing process on the magnitude levels of the vanious FRFs is shown in Fig. 15. It can be observed that the FRFs get closer to each other in terms of magnitude after balancing. Figure 15 (bottom) provides a zoomed-in view of several FRFs with different magnitude levels. It is evident that after applying the balancing technique, the lower magnitude FRFs increase in magnitude, becoming closer to those with higher initial magnitudes.

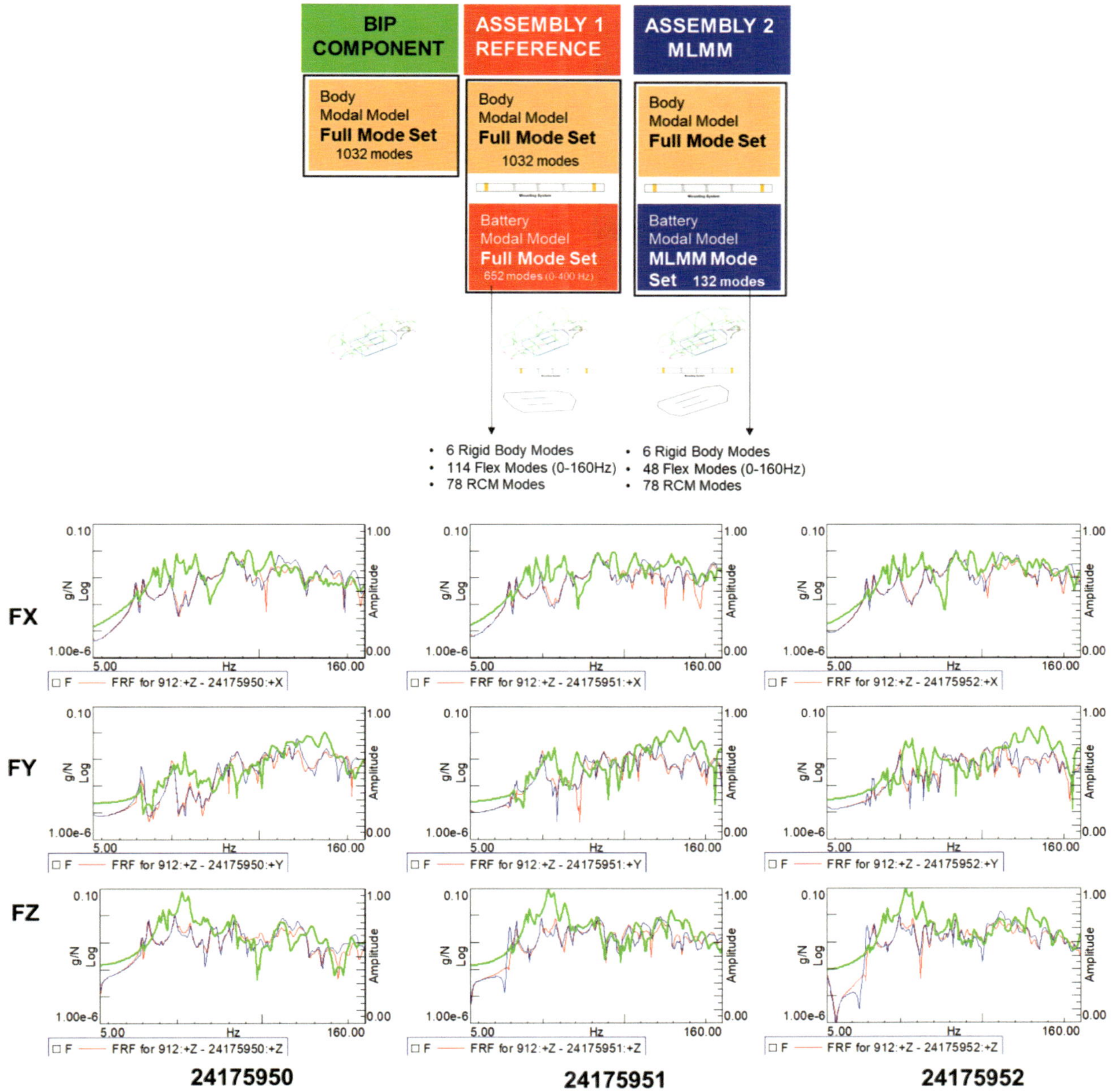

Fig. 13 Body-to-battery coupling results — Reference assembly versus the MLMM-based modal system assembly — The shown result is for the seat rail vibration as a body target response (912:+Z) and unit body load at the rear engine mount locations (24175950,24175951,24175952 +X, +Y, and +Z)

The Polymax estimator with a modal size of 200 is then applied to the balanced reduced-size FRF matrix to estimate the initial modal model, which will be used to initiate the MLMM optimization iterations. Figure 16 presents the clear stabilization diagram constructed by the Simcenter Polymax estimator, from which 47 modes were manually selected with the aid of the four Complex Mode Indicator Functions (CMIF) overlaid on the same figure. Stating with the Polymax-based initial modal model, the MLMM estimator was run for 30 iterations, imposing constraints, i.e., reciprocity, real mode shapes, and negative mass sensitivity, on the optimized modal model Fig. 17 illustrates the different iterations of the MLMM optimization process, showing a gradual decrease in the fit error over the iterations. For companison, the modal estimation process was also conducted using the imbalanced (raw) FRFs with the same user settings. Once the modal parameters are estimated they are rescaled back to make them consistent with the original magnitude level of the raw FRFs.

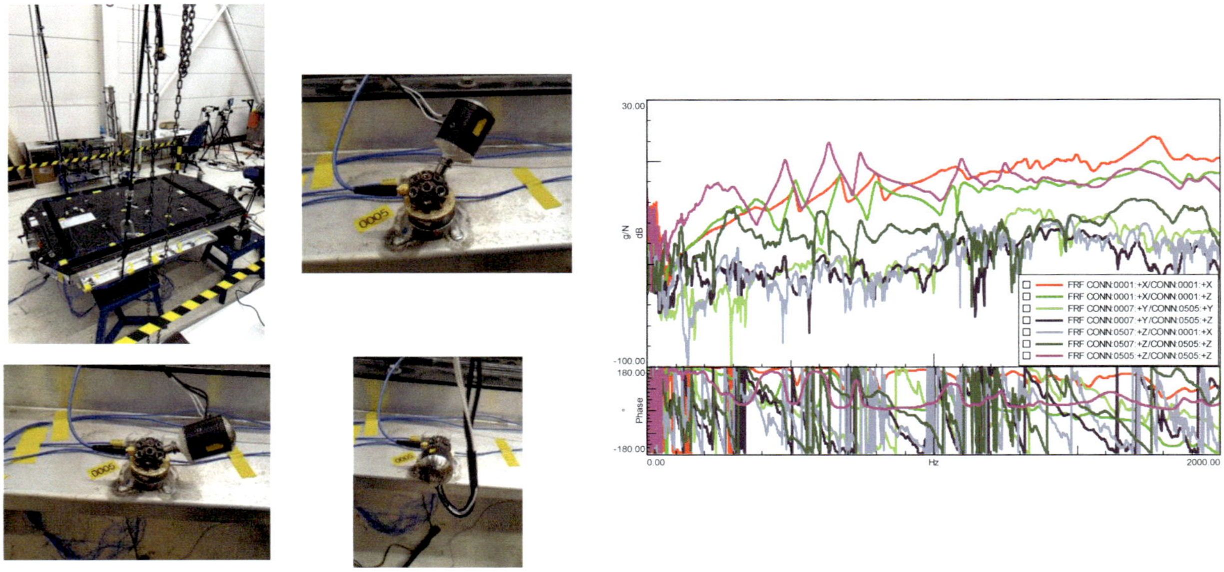

Fig. 14 The battery test set-up and some typical measured FRFs at comection points

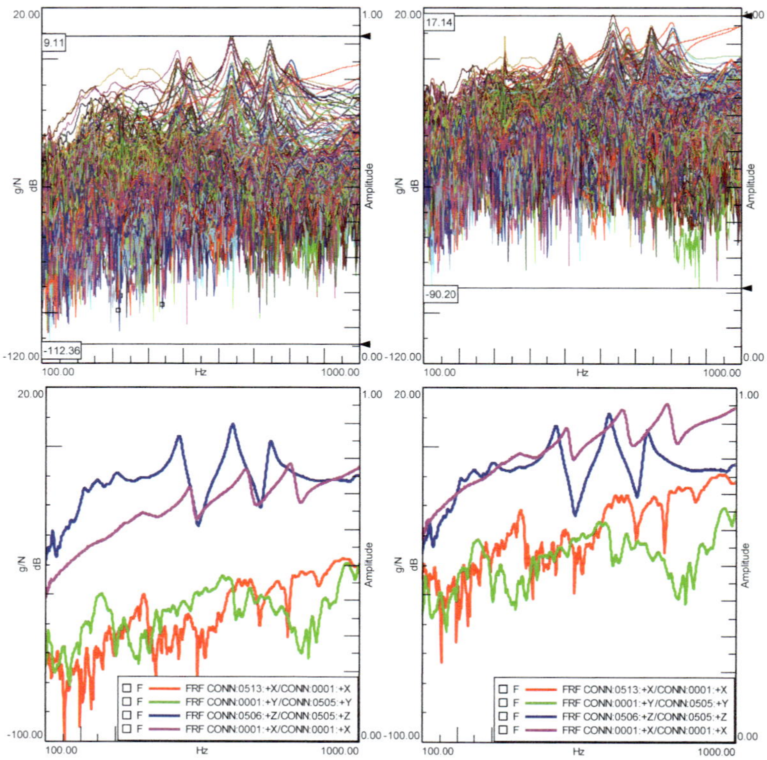

Fig. 15 The effect of the implemented FRFs balancing approach on the magnitude level of the measured FRFs (Top: all the FRFs, Bottom: zoom-in on a few FRFs, Left: imbalanced (raw) FRFs, and Right: balanced FRFs

In Fig. 18, it can be observed that for FRFs with lower initial magnitudes, the final retrieved modal synthesis is improved when the FRF matrix is balanced before performing the modal parameter estimation (see Fig. 18 - bottom). For FRFs with higher initial magnitudes (see Fig. 18 - top), the solution based on the balanced FRFs is similar to that from the raw FRFs, indicating that applying the balancing technique does not negatively affect the quality of the modal synthesis for FRFs

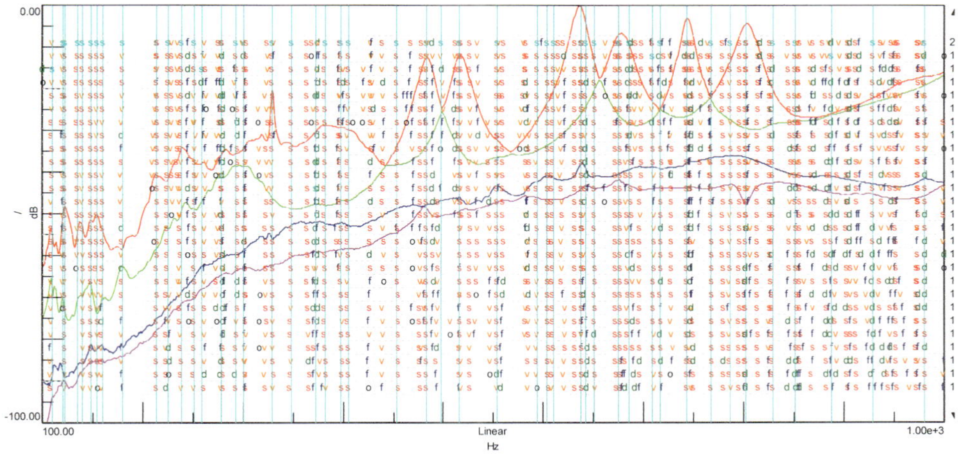

Fig. 16 Polymax stabilization chart when applied to a battery measured FRFs with 4-references (the shown curves represent the 4 Complex Mode Indicator Function CMIFs)

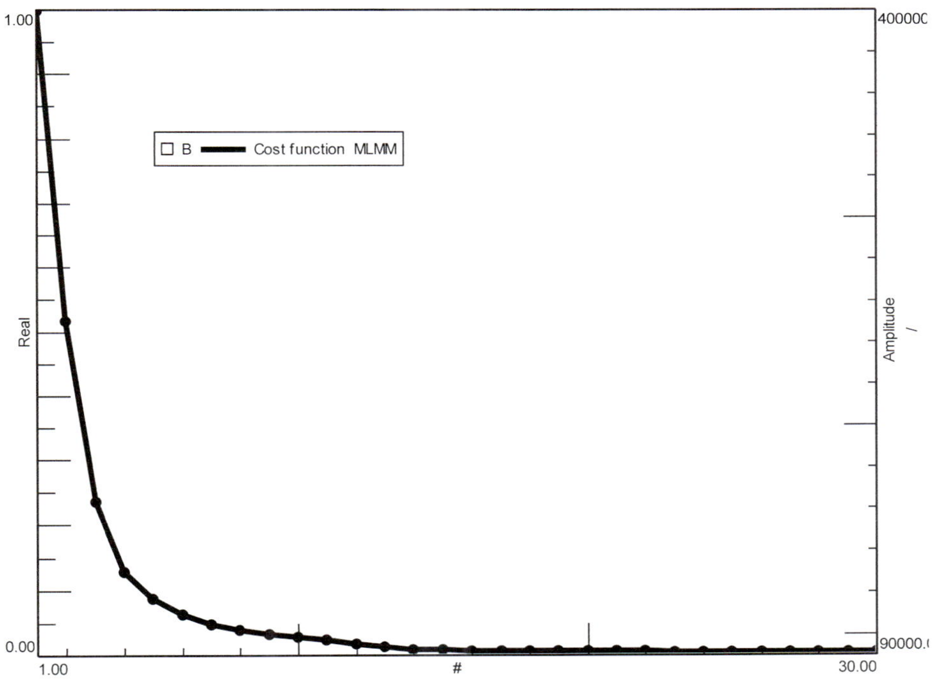

Fig. 17 The MLMM iterations to optimize a test-based battery modal model (the enror between the modal model and the true FRFs is decreasing over the different MLMM iterations)

with higher initial magnitudes. The auto-MAC of the estimated mode shapes from the two considered cases (i.e., with and without balancing the FRF matrix) is shown in Fig. 19. It is clear from the figure that using a balanced FRF matrix results in a less correlated set of mode shapes compared to the case where an unbalanced FRF matrix is used. In the final step, the reducedsize (78×4) modal model is expanded to the full size of the measured FRF matrix (78×78) by filling in the missing elements of the participation factor vectors and the diagonals of the residual term matrices. The missing participation factors for each mode are calculated using the estimated reciprocal scaling factor, thanks to the reciprocity constraint imposed in the MLMM optimization

The missing diagonal elements of the lower and upper residual term matrices are found by fitting the obtained expanded reciprocal modal model in a SISO fashion to the conesponding driving point FRF, where all the modal parameters of the

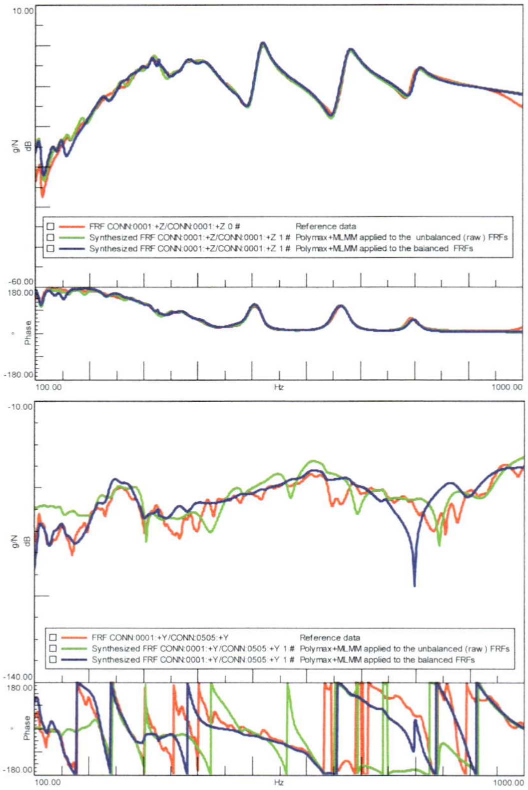

Fig. 18 Modal synthesis of the test-based MLMM estimated battery modal model (Top: FRFs with higher initial magnitudes & Bottom: FRFs with lower initial magnitudes)

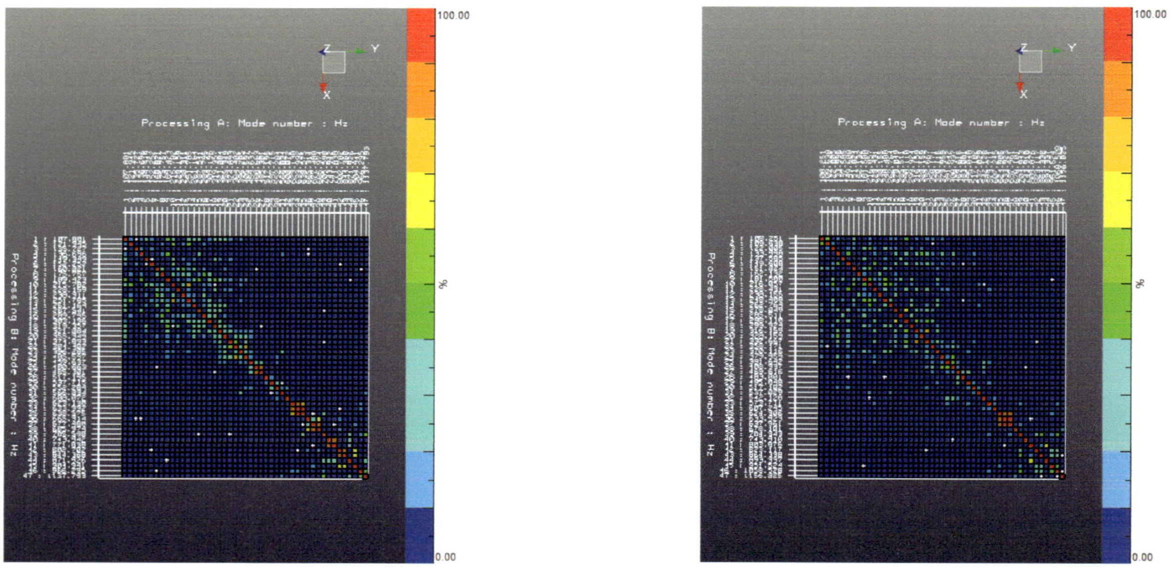

Fig. 19 The Auto-MAC function of the estimated mode shapes — Left: Estimated based on the unbalanced (raw) FRFs & Right: Estimated based on the balanced FRFs

global flexible modes are already known, and only the UR and LR terms remain as unknowns. The results of the expansion process are shown in Figs. 20 and 21, focusing on modal synthesis accuracy. In Fig. 20, the modal fit enor percentage, calculated between the measured and synthesized FRFs, is shown for both the initially considered 4 columns and the expanded

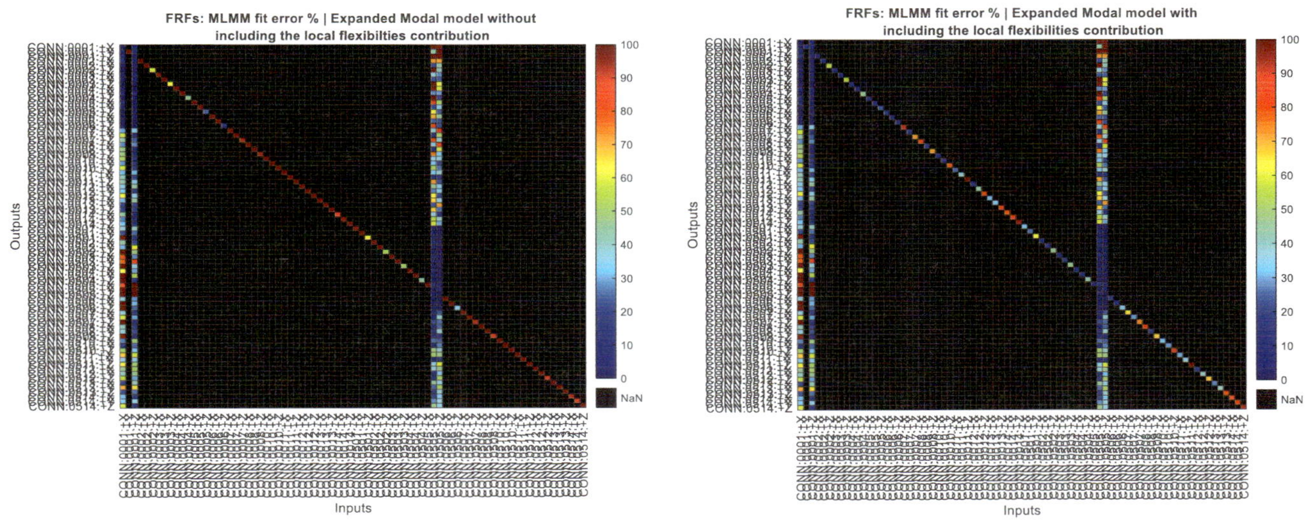

Fig. 20 Modal synthesis error % of the expanded modal model without and with including the local flexabilities contribution in the expansion (ie. upper residual term for the expanded driving points FRFs)

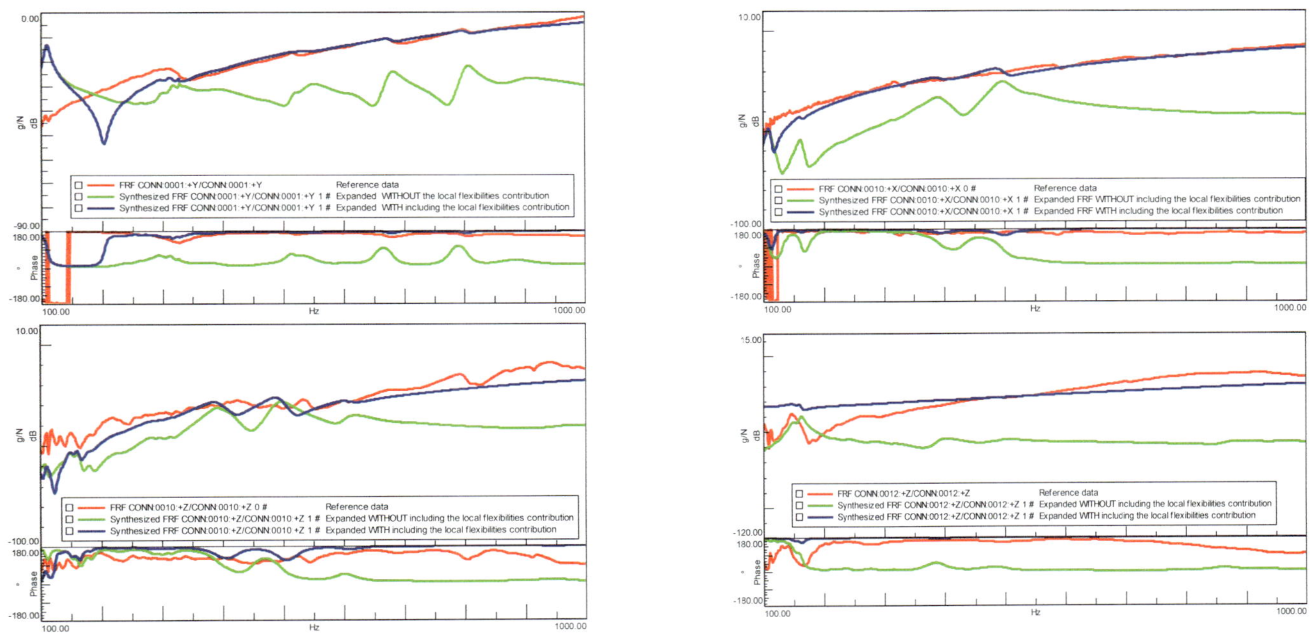

Fig. 21 Modal synthesis for driving points FRFs that were not included in the modal estimation step (i.e., FRFs calculated from the modal model expansion process)

driving point FRFs (i.e., those not included in the modal estimation step). Figure 21 illustrates the modal synthesis over the frequency band of interest for some selected expanded driving point FRFs. In general, one can see that the expansion process delivers a reasonable modal synthesis for most of the driving point FRFs that were originally not considered in the modal estimation step. The key takeaway from these figures is that expanding only the contributions of the global flexable modes is insufficient to achieve accurate synthesis for a driving point FRF that was not originally included in the modal estimation step. The inclusion of residual effects, which compensate for the trumcation of higher-frequency modes (i.e, local flexibilities), is essential for achieving an accurate modal synthesis of such an FRF.

From the presented results in this section, it can be seen that the proposed 2 -step modal parameter estimation workflow is successful in delivering a full-size modal model, acceptable in accuracy and compact in size, which considers both global flexible modes that are in within the frequency-band of interest together with the trumcated higher frequency modes' residual effects at all the considered DoFs. Key in the developed workflow is that it can be executed in a timely efficient way starting from only a reduced set of FRFs. Futhermore, the resulting modal model is directly suitable to be coupled as a modal superelement in a full vehicle (FV) Nastran model. Subsequent FV response analysis can be run efficiently in SOL111 without any frequency-dependent stiffness properties, umlike FRF-based substructuring methods.

Conclusion

In this paper, we introduced and validated an enhanced modal parameter estimation workflow aimed at improving the accuracy and efficiency of the NVH performance characterization process for electric vehicles, particularly in handling the challenges posed by large-scale unbalanced frequency response function (FRF) datasets. The workflow, leveraging the MLMM modal parameter estimator, demonstrated effectiveness in addressing common issues such as magnitude imbalances in FRF data and computational inefficiencies when dealing with a FRFs matrix with a large number of columns. Key contributions include the implementation of a magnitude-based FRF balancing technique and an expansion approach to enhance both global and local modal characteristics of the structure under test are introduced in this work. The presented magnitude-based balancing technique was proved to improve the final modal synthesis of those FRFs that come with lower initial magnitudes while not negatively affecting those which come with higher initial magnitudes. The introduced modal expansion approach helped to deliver an acceptable accurate modal model that considers the contributions of both global flexible modes and local flexibilities at all the considered DoFs in an efficient way. The findings highlight the importance of considering both global flexible modes and local flexibilities in order to have an accurate modal synthesis for all the considered DoFs, and this becomes more significant for the DoF that corresponds to a driving point. The validation, both through simulation and experimental data, confirmed that the developed workflow can efficiently handle complex fiequency response functions (FRFs) matrices, significantly reducing computational time while maintaining a reasonable accuracy in modal synthesis. Finally, the developed workflow is fit for integration in a full vehicle development process, combining seamlessly test and FE based components.

Acknowledgments The authors gratefully acknowledge the support of VLAIO (Flemish Innovation & Entrepreneurship) through the O&O project ADRENALINE (HBC.2022.1027). The authors gratefully acknowledge the support of Hyundai Motor Company (HMC) both for providing the battery hardware and the FB model.

References

1. Klerk, D. D., Rixen, D. J., & Voonmeeren, S. N. (2008). General Framework for Dynamic Substructuring: History, Review and Classification of Techniques. AIAA Joumal, 46(5), 1169-1181. doi:10.2514/1.33274
2. Peeters, B., Van der Auweraer, H., Guillaume, P., and Leuridan, J., The PolyMAX frequency-domain method: a new standard for modal parameter estimation, Shock and Vibration, vol. 11, pp. 395-409, 2004.
3. Elkafafy, M, Peeters, B., Guillaume, P., and De Troyer, T., Constrained maximum likelhood modal parameter identification applied to structural dynamics, Mechanical Systems and Signal Processing, vol. 72-73, pp. 567-589, 2016.
4. Heylen, W., Lammens, S., Sas, P., Modal Analysis Theory and Testing, Heverlee: Katholieke Universiteit Leuven, Department Werktuigkunde. 2016.
5. Lanslots, J., Rodiers, B., Peeters, B., Automated pole-selection: proof-of-concept & validation, In Proceeding of the ISMA2004 International Conference on Noise and Vibration Engineering, Leuven, Belgium, 2004.

Chapter 8
Why Don't We Just Use the Cross Spectra for Operational Modal Analysis?

William Fladung

Abstract The stochastic subspace identification algorithm, which processes correlation functions in the time domain, has been widely used as a specialized parameter estimation method for operational modal analysis for a few decades. It has also been advocated that positive power spectra, which can be derived from response-only cross spectra, can be processed by any frequency-domain parameter estimation algorithm commonly used for experimental modal analysis. This paper proposes that, with some of the currently available algorithms and modern computer speeds, estimating poles directly from cross spectra can be a viable third option.

Keywords operational modal analysis · cross spectra

Introduction

A discussion about modal parameter estimation algorithms employed in the realm of operational modal analysis (OMA) typically begins with an explanation of the frequency response function (FRF) matrix H being the relationship between the responses (or outputs) X and the forces (or inputs) F, which is then multiplied by its conjugate transpose [1, 2].

$$\{X(\omega)\} = [H(\omega)]\{F(\omega)\} \quad \rightarrow \quad \{X(\omega)\}\{X(\omega)\}^H = [H(\omega)]\{F(\omega)\}\{F(\omega)\}^H[H(\omega)]^H \tag{1}$$

With some averaging of the X and F spectra, the XX^H term becomes the output cross-spectra matrix G_{XX}, and the FF^H term becomes the input cross-spectra matrix G_{FF}. Then an assumption is made about the input power spectrum—which is not measured—being "broadband, smooth, and constant," such that it can be excluded from further consideration.

$$[G_{XX}(\omega)] = [H(\omega)][G_{FF}(\omega)][H(\omega)]^H = [H(\omega)][H(\omega)]^H \tag{2}$$

An FRF can be expressed in partial fraction form and in rational fraction polynomial form, and with the denominator polynomial in factored form to show that its roots are the poles $\lambda_r = \sigma_r + j\omega_r$, which occur in conjugate pairs, for $r = 1,\ldots,N$, where N is the number of modes.

$$H(\omega) = \sum_{r=1}^{N} \frac{A_r}{j\omega - \lambda_r} + \frac{A_r^*}{j\omega - \lambda_r^*} = \frac{\sum\limits_{k=0}^{2N-2} (j\omega)^k \beta_k}{\sum\limits_{k=0}^{2N} (j\omega)^k \alpha_k} = \frac{\sum\limits_{k=0}^{2N-2} (j\omega)^k \beta_k}{(j\omega - \lambda_1)(j\omega - \lambda_1^*)\cdots(j\omega - \lambda_N)(j\omega - \lambda_N^*)} \tag{3}$$

The denominators in the partial fraction form of the conjugate of the FRF can also be algebraically coerced into the form of jay-omega-minus-something, where that something just happens to be $-\lambda_r$.

$$H^*(\omega) = \sum_{r=1}^{N} \frac{A_r^*}{(j\omega - \lambda_r)^*} + \frac{(A_r^*)^*}{(j\omega - \lambda_r^*)^*} = \sum_{r=1}^{N} \frac{A_r^*}{-j\omega - \lambda_r^*} + \frac{A_r}{-j\omega - \lambda_r} = \sum_{r=1}^{N} \frac{-A_r}{j\omega - (-\lambda_r)} + \frac{-A_r^*}{j\omega - (-\lambda_r^*)} \tag{4}$$

William Fladung
ATA Engineering, Inc., San Diego, California 92128, USA
e-mail: bill.fladung@ata-e.com

Brandon Dilworth et al. (eds.), *Topics in Modal Analysis & Parameter Identification II, Vol. 10*, Conference Proceedings of the Society for Experimental Mechanics Series, https://doi.org/10.13052/9788743801559-8

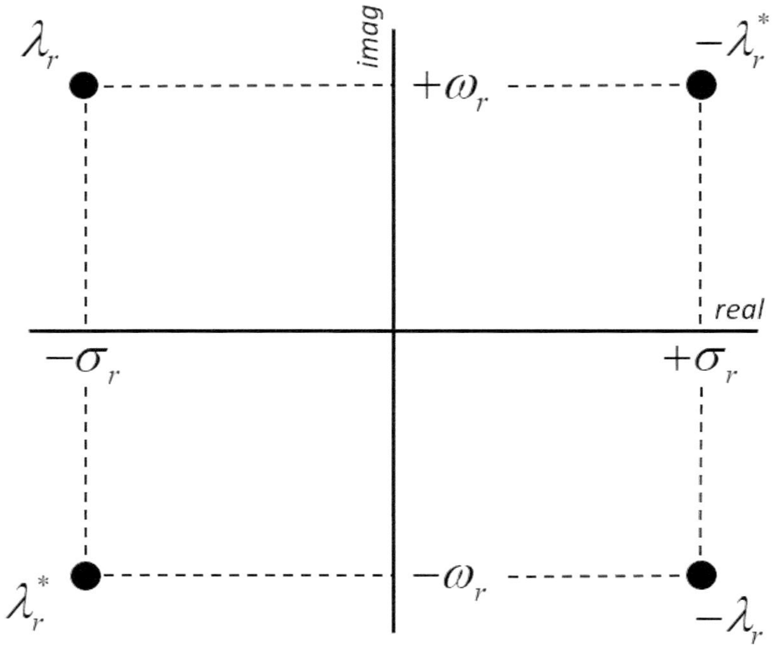

Fig. 1 The quartet of poles contained in a cross spectrum on the complex plane

Whereas an FRF contains pairs of conjugate poles, it can be seen from Eqs. (3) and (4) that a cross spectrum contains quartets of poles, with conjugate pairs of the *positive* poles (λ_r) in the left half plane and conjugate pairs of the *negative* poles ($-\lambda_r$) in the right half plane, as shown in Fig. 1.

Combining H and H^* results in a partial fraction form for a cross spectrum G_{xx} that contains four terms, with two for the positive poles and two for the negative poles. However, the constants in the numerators are no longer true residues A_r because they do not contain modal scaling since the force is not measured.

$$G_{xx}(\omega) = \sum_{r=1}^{N} \frac{R_r}{j\omega - \lambda_r} + \frac{R_r^*}{j\omega - \lambda_r^*} - \frac{\bar{R}_r}{j\omega - (-\lambda_r)} - \frac{\bar{R}_r^*}{j\omega - (-\lambda_r^*)} \tag{5}$$

The inverse Fourier transform of a cross spectrum is a cross correlation, as depicted in Fig. 2a and b. The exponentially decaying response in the positive-lags section of the correlation function corresponds to the positive poles, and the exponentially growing response in the negative-lags section corresponds to the negative poles. A popular OMA method for estimating the poles from the positive-lag section of the correlations is the stochastic subspace identification (SSI) [3] algorithm, but any other time-domain method, such as the polyreference time domain (PTD) [4] algorithm, could also be used.

Alternatively, a *positive power spectrum* (PPS) [1], or *half spectrum* [2], can be created from the Fourier transform of the positive-lags section of the correlation function, as depicted in Fig. 2c and d. Since a PPS contains only the two positive frequency terms in Eq. (5), it has the same partial fraction form as an FRF, and any frequency-domain algorithm can be used to estimate the poles. However, the PPS has a few peculiar traits. The first is that its phase crosses 0° or 180° at resonance (as denoted by the blue markers in Fig. 2d) instead of ±90° as does the phase of an acceleration/force FRF, and all PPSs have this same phase behavior regardless of the response data type. In addition, the act of creating a PPS tends to smooth the function, which often obscures some of the lower peaks, as illustrated in Fig. 3. Truncating a cross correlation to the positive lags (or zeroing the negative lags) has the same effect as multiplying it by a unit step function [1], the Fourier transform of which is $\pi\delta(\omega) + 1/j\omega$ and is shown in the Fig. 3 inset. Convolution of the $1/j\omega$ term with the cross spectrum seems to fill in the valleys between the peaks. This behavior is also evident by comparing Fig. 2a and d.

Furthermore, there is always the possibility of introducing truncation errors when applying the discrete Fourier transform, both when going from the frequency domain to the time domain to produce correlations from cross spectra and again when going back to the frequency domain with the PPS. The former, which would affect time-domain methods such as SSI or PTD, could be avoided by computing correlations in the time domain (which could be time consuming) or by using other signal processing techniques [5]. The latter would be leakage if the correlation response does not decay to zero in the frame of pos-

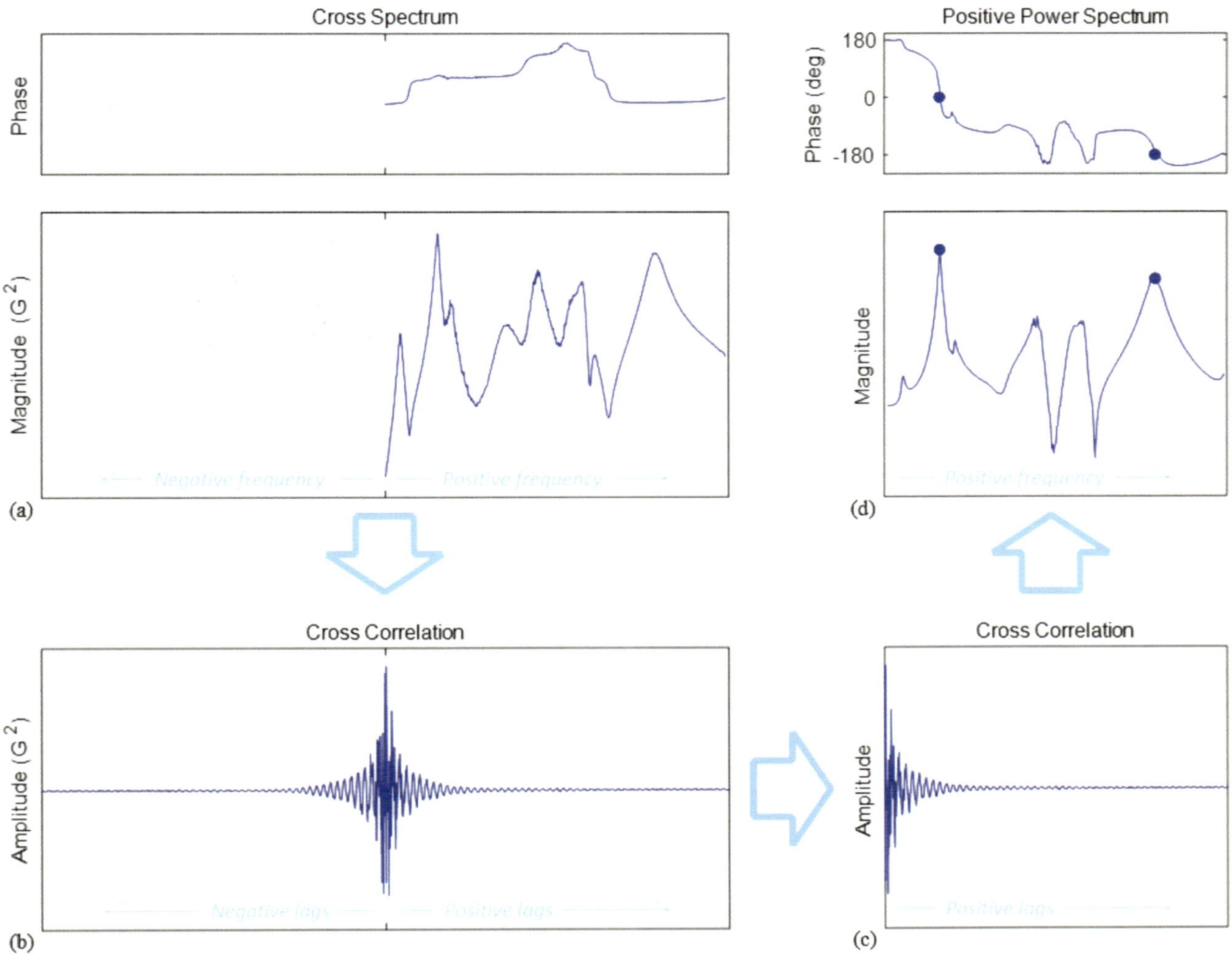

Fig. 2 (a) A double-sided acceleration-to-acceleration cross spectrum, (b) the cross correlation of the cross spectrum in (a), (c) the positive lags of the cross correlation in (b), and (d) the positive power spectrum from the cross-correlation positive lags in (c)

itive lags. Applying an exponential window to the correlations can reduce the effects of leakage [2] but at the expense of artificially increasing the damping, which has the effect of further smoothing the PPS and possibly further obscuring the peaks.

Using Cross Spectra for Estimating Poles

Which brings us to the question posed in the title of this paper. The rational fraction polynomial form of the FRF was given in Eq. (3), and a similar expression for its conjugate is given in Eq. (6), with different (i.e., primed) alpha and beta coefficients and conjugate pairs of the negative poles in the factored form of the denominator polynomial.

$$H^*(\omega) = \frac{\sum_{k=0}^{2N-2} (j\omega)^k \beta'_k}{\sum_{k=0}^{2N} (j\omega)^k \alpha'_k} = \frac{\sum_{k=0}^{2N-2} (j\omega)^k \beta'_k}{(j\omega - (-\lambda_1))(j\omega - (-\lambda_1^*)) \cdots (j\omega - (-\lambda_N))(j\omega - (-\lambda_N^*))} \tag{6}$$

Combining H and H^* to form the cross spectrum as $G_{xx} = HH^*$ produces a rational fraction polynomial model in Eq. (7), with again different (i.e., hatted) alpha and beta coefficients and a characteristic equation in the denominator that has both

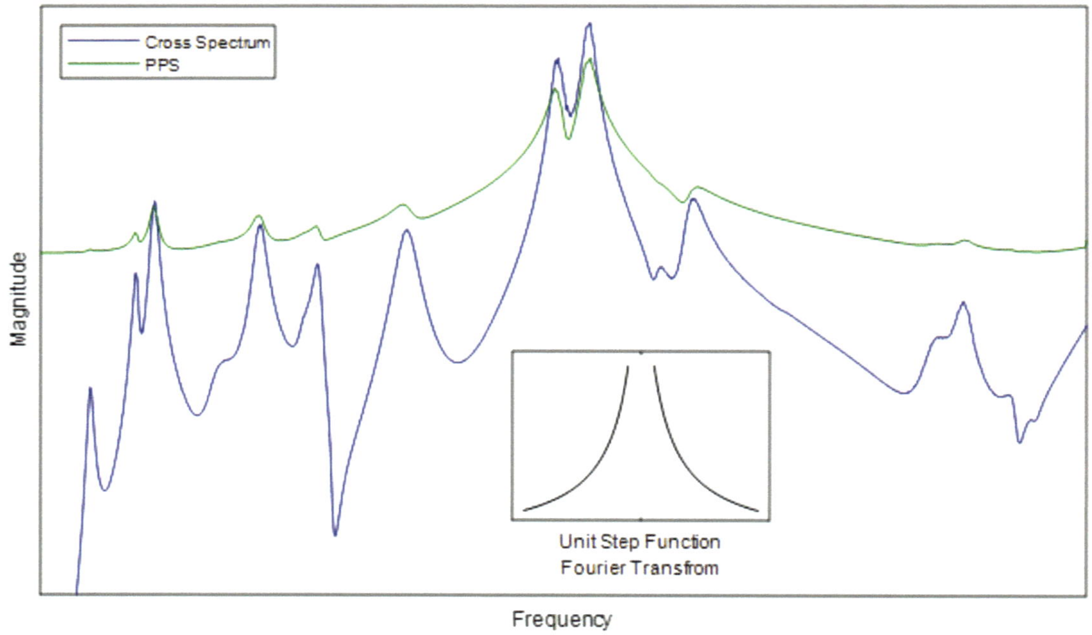

Fig. 3 A cross spectrum and its positive power spectrum

the positive poles (and their conjugates) and the negative poles (and their conjugates) as its roots. The model order of the denominator polynomial is now twice that of an FRF, and the order of the numerator polynomial is 4 less than the order of the denominator polynomial instead of 2.

$$G_{xx} = \frac{\sum_{k=0}^{4N-4} (j\omega)^k \hat{\beta}_k}{(j\omega - \lambda_1)(j\omega - \lambda_1^*)\cdots(j\omega - (-\lambda_1))(j\omega - (-\lambda_1^*))\cdots(j\omega - (-\lambda_N))(j\omega - (-\lambda_N^*))} = \frac{\sum_{k=0}^{4N-4} (j\omega)^k \hat{\beta}_k}{\sum_{k=0}^{4N} (j\omega)^k \hat{\alpha}_k} \tag{7}$$

Equation (7) forms the basis for extending the rational fraction polynomial (RFP) [6] algorithm to use cross spectra instead of FRFs. A least-squares solution is formulated for the unknown alphas and betas, and an eigenvalue solution of a companion matrix formed with the alphas yields the poles. However, it is well known that the RFP algorithm inevitably becomes ill-conditioned as the frequencies are raised to higher powers, which will only be exacerbated when the model order is doubled. But while this may be true when using high-order models with power polynomials, the ill-conditioning predicament can be evaded by using a different algorithm, such as the Laplace-domain orthogonal polynomial (LDOP) [7] algorithm or the Z-domain orthogonal polynomial (ZDOP) [8] algorithm. These two algorithms are similar to their ancestor RFP in that they are high-order models with the alpha coefficients being the size of the number of references—but instead of power polynomials, their basis functions are orthogonal polynomials (in either the Laplace or Z domain), which have favorable numerical conditioning characteristics regardless of the model order. In addition, the polyreference least-squares complex frequency (PLSCF) [9] and the z-domain rational fraction polynomial (RFP-Z) [10] algorithms use a complex frequency mapping to the z-domain to forgo the use of power polynomials, making them viable candidates for using cross spectra for OMA as well. A low-order frequency-domain algorithm [10], which has coefficient matrices the size of the number of responses and a model order of 2, could also be adapted for OMA with cross spectra by using a model order of 4.

For a modal test (i.e., the kind where the excitation is supplied by shakers and the forces are measured), the number of references is typically less than about ten and much less than the number of responses, which could be as many as a few hundred. For an OMA test, there could also be hundreds of response channels measured. The $N_o \times N_o$ output cross-spectra matrix $[G_{XX}]$ in Eq. (2) is created (at each frequency) by the outer product of the $N_o \times 1$ vector of responses $\{X\}$ from Eq. (1), which means that there would be N_o references, too. Obviously, using hundreds of references in a high-order algorithm such as RFP, ZDOP, LDOP, PLSCF, or RFP-Z would generate an impractically large number of poles as the model order increases. While the response channels could be down-selected to a more reasonable number of references, the criteria

Table 1 Time trials of ZDOP solutions using PPS and cross spectra

Spectral Lines	Responses	References	Minimum Order	Step Order	Maximum Order	PPS	Cross Spectra
1600	100	5	10	2	40	3 s	9 s
1600	100	10	5	1	20	3 s	10 s
1600	200	5	10	2	40	5 s	15 s
1600	200	10	5	1	20	5 s	17 s

for doing so may not be overly intuitive. Alternatively, the reference space could be condensed to a few *virtual references* using a subset of vectors from the singular value decomposition of a concatenated array of G_{XX} at all spectral lines over the frequency range of interest.

Some preemptive time trials were run for a few feasibly large datasets, with the results listed in Table 1. The number of poles produced from the characteristic equation is equal to the model order times the size of the coefficient matrices, which for a high-order model such as ZDOP is the number of references. The model orders listed in Table 1 correspond $2N$ in the denominator polynomials in Eqs. (3) and (6), and the model orders for the cross spectra were double those listed. The number of spectral lines corresponds to the originating cross spectra and the PPS have half that many (and double the frequency spacing) after the negative lags are truncated. Unsurprisingly, solving for twice the model order takes longer to complete—three or more times longer from a relative point of view. But in actual time that somebody would have to wait, it's only a few more seconds.

As a proof of concept, poles were estimated from a random excitation test run on an "iron bird" laboratory test article, shown in Fig. 4a. There was a vertical shaker attached to both of the wings and horizontal stabilizers, as indicated by the arrows, and 109 responses. Cross spectra were generated from the FRFs as HH^H, and PPSs were generated from the cross spectra. Although this is admittedly not an OMA test, $[G_{FF}]$ in Eq. (2) is truly an identity matrix in this case, and the stability diagram for the FRFs in Fig. 4b serves as the baseline for evaluating the PPS and cross-spectra results. The gold stars in the top row are autonomously selected poles [11–13], and the filled blue squares are manually selected poles. The stability diagram for PPSs is shown in Fig. 5a, and the stability diagram for cross spectra is shown in Fig. 5b. The model orders on the y-axis of the stability diagrams correspond to the positive poles of the underlying FRF model (i.e., $2N$ in the denominator polynomials in Eqs. (3) and (6)), and the model orders for the cross-spectra solutions were double those. Using the ZDOP algorithm with zeroth-order coefficient normalization yields clear stability diagrams [14] for all three, with quite

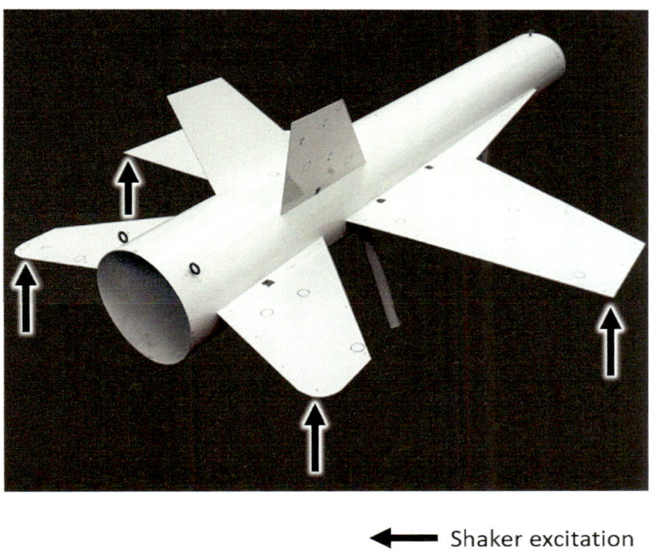

(a)

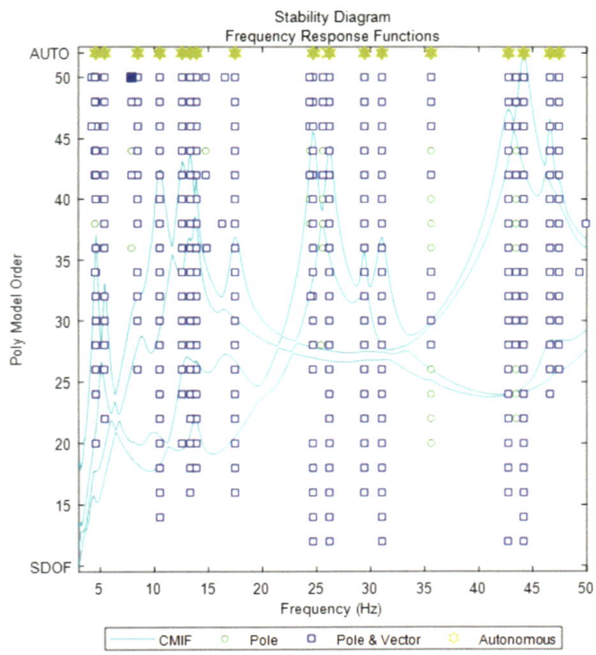

(b)

Fig. 4 (a) The iron bird laboratory test article; (b) ZDOP stability diagram for FRFs

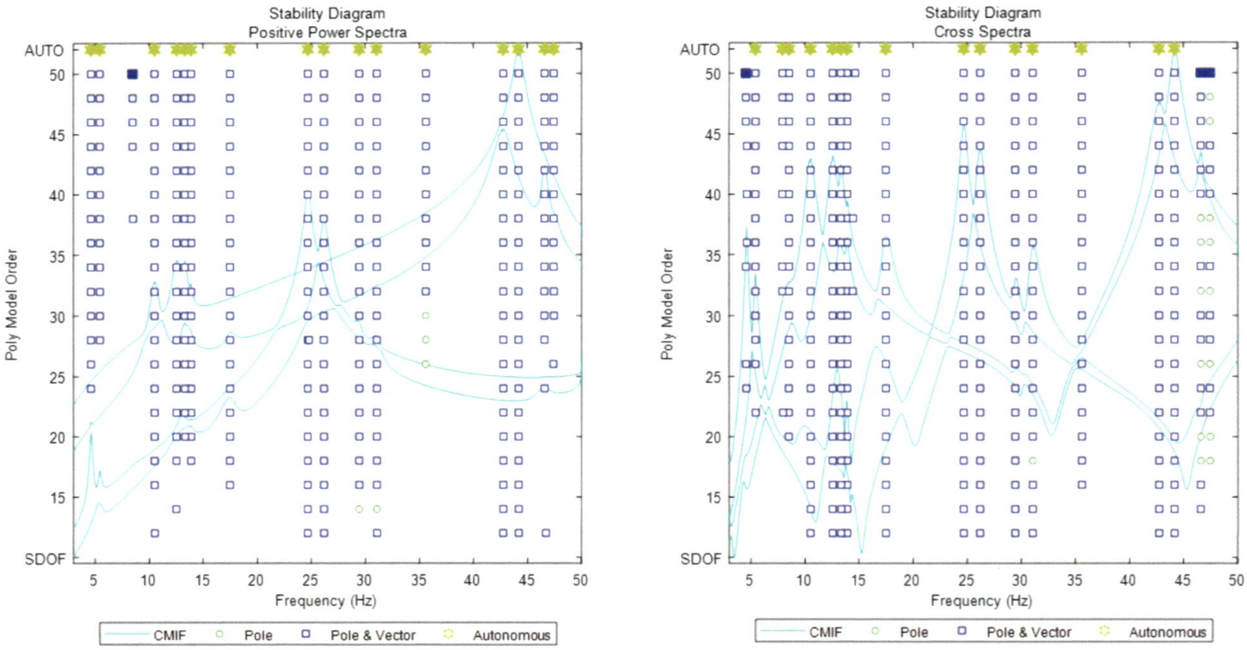

(a) (b)

Fig. 5 (a) ZDOP stability diagram for PPS; (b) ZDOP stability diagram for cross spectra

satisfactory and generally comparable results. One exception is that while a few consistent poles at 8.5 Hz were found with PPS for higher model orders, the mode at 7.9 Hz was missed entirely, and there's no evidence of either in the mode indicator functions. The frequency and damping values for the selected poles from the three stability diagrams are listed in Table 2, which verifies that the frequencies match well, albeit with some disagreement about the damping for a few of the modes.

Table 2 Selected poles from stability diagrams in Fig. 4b and Fig. 5

	FRF		PPS		Cross Spectra	
Mode No.	**Frequency (Hz)**	**Damping (%)**	**Frequency (Hz)**	**Damping (%)**	**Frequency (Hz)**	**Damping (%)**
1	4.59	2.04	4.60	2.19	4.55	3.02
2	5.39	3.37	5.42	3.53	5.42	4.33
3	7.88	8.51	—	—	7.92	7.38
4	8.46	7.90	8.46	7.96	8.52	8.06
5	10.49	3.55	10.48	3.59	10.50	3.58
6	12.54	3.03	12.54	2.92	12.55	2.99
7	13.32	1.75	13.31	1.62	13.31	1.62
8	13.88	0.92	13.88	0.90	13.88	0.92
9	17.46	3.04	17.47	3.02	17.48	3.04
10	24.67	1.18	24.67	1.17	24.67	1.12
11	26.18	1.29	26.17	1.27	26.18	1.25
12	29.45	1.03	29.45	1.02	29.45	1.03
13	31.05	1.53	31.05	1.53	31.05	1.53
14	35.58	1.30	35.58	1.30	35.58	1.30
15	42.76	1.63	42.76	1.58	42.73	1.62
16	44.18	0.98	44.17	0.96	44.16	0.95
17	46.60	0.53	46.58	0.52	46.59	0.50
18	47.42	0.99	47.42	0.76	47.43	0.85

Using Cross Spectra for Estimating Residues

It should come as no surprise that the partial fraction model for residue estimation has four terms for each mode. Modifying Eq. (5) by collecting the terms in the summation and including residuals gives the single-reference residues estimation formulation as

$$G_{xx}(\omega) = \sum_{r=1}^{4N} \frac{R'}{j\omega - \lambda_r} + \left(\frac{1}{\omega^2}\right)^2 \rho_L + (1)^2 \rho_U, \tag{8}$$

where 4*N* in the upper limit of the summation signifies quartets of poles, *R'* (i.e., prime) stands in for both variants of the *R*'s and their conjugates in Eq. (5), λ_r can be either a positive or negative pole, ρ_L is the lower residuals, and ρ_U is the upper residuals. The residual terms, which are derived from the asymptotic behavior of a single degree of freedom FRF far away from its resonance, need to be squared since the cross spectrum is defined from HH^H. A least-squares solution for the residues and residuals is generated by rearranging Eq. (8) into matrix form and evaluating at different frequencies.

$$\begin{bmatrix} G_{xx}(\omega_1) \\ G_{xx}(\omega_2) \\ G_{xx}(\omega_3) \\ \vdots \end{bmatrix} = \left[\begin{array}{ccc|cc} \Lambda_1(\omega_1) & \cdots & \Lambda_{4N}(\omega_1) & \omega_1^{-4} & 1 \\ \Lambda_1(\omega_2) & \cdots & \Lambda_{4N}(\omega_2) & \omega_2^{-4} & 1 \\ \Lambda_1(\omega_3) & \cdots & \Lambda_{4N}(\omega_3) & \omega_3^{-4} & 1 \\ \vdots & & \vdots & \vdots & \vdots \end{array}\right] \begin{bmatrix} R'_1 \\ \vdots \\ R'_{4N} \\ \hline \rho_L \\ \rho_U \end{bmatrix}, \quad \text{where } \Lambda_r(\omega) = \frac{1}{j\omega - \lambda_r} \tag{9}$$

In general, the residues are (or can be) complex, but a phase-constrained least-squares solution for real (or normal) mode shapes [15] can be formulated by partitioning the $Ax = b$ form of Eq. (9) as

$$\hat{A}\hat{x} + c_L\rho_L + c_U\rho_U = b, \quad \text{where } \hat{x} = \begin{bmatrix} R'_1 & \cdots & R'_{4N} \end{bmatrix}^T, c_L = \omega^{-4}, c_U = 1, \tag{10}$$

$\hat{A}$ is the left partition of *A*, *b* is the cross spectrum, and ρ_L and ρ_U are real such that the residual terms are purely real. Then expand the complex terms into their real and imaginary parts as

$$\left(\hat{A}_r + j\hat{A}_i\right)(\hat{x}_r + j\hat{x}_i) + c_L\rho_L + c_U\rho_U = b_r + jb_i, \quad \text{where } \hat{A}_r = \mathrm{Re}\left(\hat{A}\right), \hat{A}_i = \mathrm{Im}\left(\hat{A}\right), \tag{11}$$

and likewise for $\hat{x}$ and *b*. Next, expand the product of the two complex terms and collect the real and imaginary components into separate equations as

$$\begin{aligned} \hat{A}_r\hat{x}_r - \hat{A}_i\hat{x}_i + c_L\rho_L + c_U\rho_U &= b_r \\ \hat{A}_i\hat{x}_r + \hat{A}_r\hat{x}_i &= b_i \end{aligned} \tag{12}$$

The response and reference (i.e., force) of an FRF are 90° out of phase at resonance and the residue is purely imaginary, which means $\hat{x}_r = 0$ for real mode shapes. However, response and reference of a cross spectrum are really both responses (e.g., acceleration) and must be either in phase (i.e., 0°) or out of phase (i.e., 180°) for a real mode shape, which means that the residue must be purely real and $\hat{x}_i = 0$. Removing those terms and combining into matrix form gives the normal mode solution as

$$\begin{bmatrix} \hat{A}_r & c_L & c_U \\ \hat{A}_i & 0 & 0 \end{bmatrix} \begin{Bmatrix} \hat{x}_r \\ \rho_L \\ \rho_U \end{Bmatrix} = \begin{Bmatrix} b_r \\ b_i \end{Bmatrix} \tag{13}$$

The above alterations to the residues algorithm (i.e., four terms in the summation, squaring of the residuals, and real shape coefficients) are also applicable to the multiple-reference formulation.

Figure 6 shows the complex mode indicator function (CMIF) computed from the measured cross spectra (b) and from the PPS derived from the measured cross spectra (a). Also plotted are the CMIFs computed from the cross spectra and PPS synthesized with the poles in Table 2 and estimated residues. For both cases, the synthesized functions match well with the measured functions at the prominent peaks but not as well at the less well excited (and less well indicated) peaks. The modal assurance criteria (MAC) between shapes from the FRFs and from the PPS and cross spectra, shown in Fig. 7, reach the same consensus, with the shapes matching well for well excited (and well indicated) modes but less so otherwise.

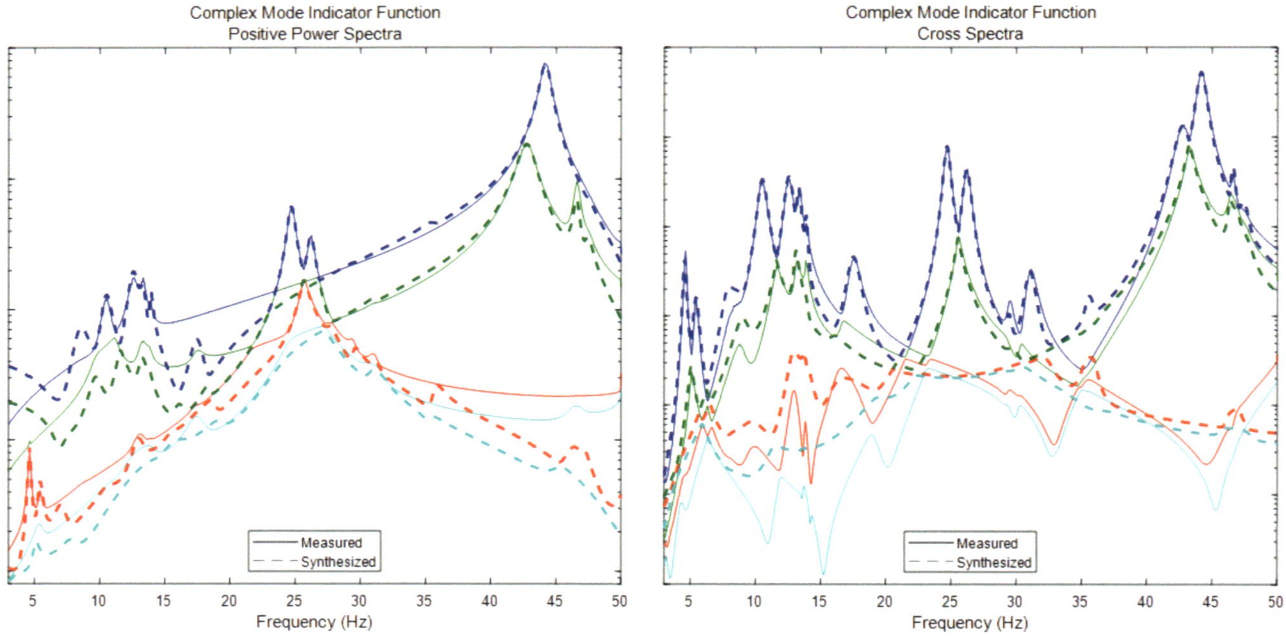

(a) (b)

Fig. 6 (a) CMIF from measured and synthesized PPS; (b) CMIF from measured and synthesized cross spectra

		Cross MAC Table FRF Shapes																	
		1	2	3	4	5	6	7	8	9	10	11	12	13	14	15	16	17	18
PPS Shapes		4.59	5.40	7.91	8.49	10.49	12.55	13.32	13.88	17.47	24.67	26.18	29.45	31.05	35.58	42.76	44.18	46.60	47.43
1	4.61	86									29								
2	5.43		59								28	29							
3	8.49																96		
4	10.49					95										36			
5	12.54				29		85										45		
6	13.31							90											
7	13.88							24									73		
8	17.48					23				34						69			
9	24.67										100								
10	26.17											100							
11	29.45																75		
12	31.05					25								62		44			
13	35.58																95		
14	42.76															99			
15	44.17																100		
16	46.58																	91	27
17	47.42															25		36	

(a)

		Cross MAC Table FRF Shapes																	
		1	2	3	4	5	6	7	8	9	10	11	12	13	14	15	16	17	18
XSPEC Shapes		4.59	5.40	7.91	8.49	10.49	12.55	13.32	13.88	17.47	24.67	26.18	29.45	31.05	35.58	42.76	44.18	46.60	47.43
1	4.56	98																	
2	5.42		72																
3	7.94															40			
4	8.55				51												34		
5	10.51					99													
6	12.55				27		97												
7	13.31							98											
8	13.88							78	32										
9	17.49					24				85									
10	24.67										100								
11	26.18											100							
12	29.45											54							
13	31.05													88					
14	35.58															47	44		
15	42.74															99			
16	44.16																100		
17	46.59																	96	31
18	47.44																	44	

(b)

Fig. 7 (a) Cross-MAC for FRF shapes vs PPS shapes; (b) cross-MAC for FRF shapes vs cross-spectra shapes

Missing Phase Crossings at Resonance

Figure 8a shows two FRFs from the wingtips of another mock aircraft test article (shown in the inset). The drive point FRF exhibits the expected phase characteristics with drops through 90° at resonance (indicated by the black markers). The FRF for the opposite wingtip also has phase drops through ±90° at resonance, with the first and fourth modes in phase and second and third modes out of phase. Figure 8b shows the cross spectra for the responses of the FRFs in Fig. 8a, generated from HH^H. The cross spectrum for the drive point response, which is actually an auto spectrum, has phase of zero at all frequencies. The phase of the cross spectrum for the opposite wingtip is again in phase (i.e., −360°) for the first and fourth modes and out of phase (i.e., −180°) for the second and third modes. Thus, one compromise inherent to using cross spectra for OMA is the missing phase crossings at resonance, which can sometimes offer another opinion when evaluating the estimated poles. For example, the phase crossing can verify the undamped natural frequency and the slope of the phase can assess the damping of a heavily damped mode that may not have a distinct peak indicating the resonance.

As a side effect of cross spectra not having ±90° phase crossings at resonance, the normal mode indicator function (NMIF) and the multivariate mode indicator function (MMIF) cannot be used with cross spectra. This is because the NMIF is derived from the ratio of the real part to the magnitude of FRFs, and the drops in the functions occur when the real part is

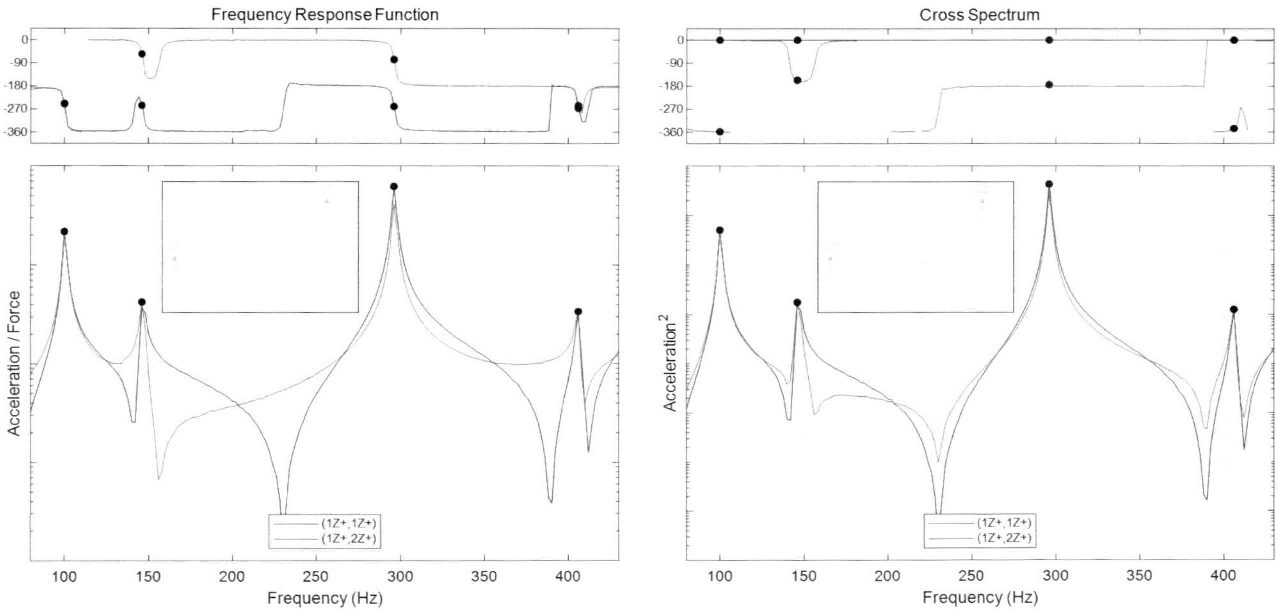

(a) (b)

Fig. 8 (a) Two mock airplane wingtip FRFs; (b) cross spectra for responses of the FRFs in (a)

zero, which is the same as a phase of 90°. The same reason applies to the MMIF, since it is the multiple-reference extension of the NMIF [16].

Summary

Whereas correlation functions and PPS have conventionally been used for OMA parameter estimation, this paper has shown that, with a prudent choice of an algorithm and a bit of patience, so too can cross spectra. A quartet of positive and negative poles is estimated from cross spectra, requiring twice the model order and more than twice the time. But while the computational burden might have been an obstacle in the past, today the additional wait time should only be a few seconds. The residue estimation formulation is also adaptable to cross spectra for either complex or real shapes. The benefits of using cross spectra are being able to use standard spectral estimation methods and avoiding any potentially error-inducing trips between the temporal domains. Of course, there are necessarily trade-offs with using cross spectra, such as not having ±90° phase crossings at resonance and not being able use certain mode indicator functions.

References

1. Chauhan, S., Martell, R., Allemang, R.J., and Brown. D.L., "Unified Matrix Polynomial Approach for Operational Modal Analysis," *Proceedings of the 25th International Modal Analysis Conference*, February 2007.
2. Peeters, B., and Van der Auweraer, H., "PolyMax: a revolution in operational modal analysis," *Proceedings of the 1st International Operational Modal Analysis Conference*, April 2005.
3. Chauhan, S., "Using the Unified Matrix Polynomial Approach (UMPA) for the Development of the Stochastic Subspace Identification (SSI) Algorithm," *Journal of Vibration and Control*, Vol. 19, No. 13, pp. 1950–61, October 2013.
4. Vold, H., Kundrat, J., Rocklin, T., and Russell, R., "A Multi-Input Modal Estimation Algorithm for Mini-Computers," *SAE Transactions*, Vol. 91, No. 1, pp. 815–21, January 1982.
5. Xu, Y.F., Liu, J.M., and Zhu, W.D., "Accurate and Efficient Calculation of Discrete Correlation Functions and Power Spectra," *Journal of Sound and Vibration*, Vol. 347, pp. 246–65, July 2015.
6. Richardson, M., and Formenti, D.L., "Parameter Estimation from Frequency Response Measurements Using Rational Fraction Polynomials," *Proceedings of the 1st International Modal Analysis Conference*, 1982.

7. Fladung, W.,and H. Vold, "An Orthogonal View of the Polyreference Least-Squares Complex Frequency Modal Parameter Estimation Algorithm," *Proceedings of the 33rd International Modal Analysis Conference*, February 2015.
8. Fladung, W., and H. Vold, "An Improved Implementation of the Orthogonal Polynomial Modal Parameter Estimation Algorithm Using the Orthogonal Complement," *Proceedings of the 33rd International Modal Analysis Conference*, February 2015.
9. Guillaume, P., Verboven, P., Vanlanduit, S., Van der Auweraer, H., and Peeters, B., "A Poly-Reference Implementation of the Least-Squares Complex Frequency-Domain Estimator," *Proceedings of the 21st International Modal Analysis Conference*, February 2003.
10. Chauhan, S., Martell, R., Allemang, R.J., and Brown. D.L., "Implementation of Complex Frequency Mapping to Low Order Frequency Domain algorithm for Operational Modal Analysis," *Proceedings of the 25th International Modal Analysis Conference*, February 2007.
11. Phillips, A.W., Allemang, R.J., and Brown, D.L., "Autonomous Modal Parameter Estimation: Methodology," *Proceedings of the 29th International Modal Analysis Conference*, February 2011.
12. Phillips, A.W., Allemang, R.J., and Brown, D.L., "Autonomous Modal Parameter Estimation: Statistical Considerations," *Proceedings of the 29th International Modal Analysis Conference*, February 2011.
13. Phillips, A.W., Allemang, R.J., and Brown, D.L., "Autonomous Modal Parameter Estimation: Application Examples," *Proceedings of the 29th International Modal Analysis Conference*, February 2011.
14. Verboven, P., Cauberghe, B., Vanlanduit, S., Parloo, E., and Guillaume, P., "The Secret Behind Clear Stabilization Diagrams: The Influence of the Parameter Constraint on the Stability of the Poles," *Proceedings, Society of Experimental Mechanics (SEM) Annual Conference*, 2004.
15. Fladung, W., and Napolitano, K., "Tips, Tricks, and Obscure Features for Modal Parameter Estimation," *Proceedings of the 41st International Modal Analysis Conference*, February 2023.
16. Williams, R., Crowley, J., Vold, H., "The Multivariate Mode Indicator Function in Modal Analysis," *Proceedings of the 3rd International* Modal Analysis Conference, 1985.

Chapter 9
FRF Error Mitigation Methods for time Waveform Replication

Keaton Coletti, Steven Carter, and Ryan Schultz

Abstract Time waveform replication (TWR) must be performed carefully to avoid errors in the output waveform. A key source of error in multi-input, multi-output TWR is error in the measured frequency response functions (FRFs). If an inverse Fast Fourier Transform (IFFT) is performed on the measured FRFs, the error appears as non-causal growth at the end of the resulting signals. When non-causal FRFs are applied to TWR control, significant errors between the target and system responses occur: in particular, the system responds to transient events such as shocks before prescribed in the target response. This paper shows that non-causal effects due to noise are distinct from effects due to finite bandwidth in measured FRFs. Using a TWR framework that does not require any time-domain windowing, several FRF error mitigation techniques are explored. The techniques are compared in a time waveform replication simulation featuring three inputs and eight outputs.

Keywords time-waveform replication · multi-input multi-output · FRF causality · transient control

Introduction

Time waveform replication (TWR) is the process of replicating a time-domain waveform in the output of a dynamic system. In structural dynamics, performing TWR in a laboratory setting is critical to the qualification of, *e.g.*, automotive and aeronatical components and assemblies. These may be subjected to complicated loading environments during service, including combinations of random vibration, harmonic vibration, transient vibration, and shock. Multi-input, multi-output (MIMO) testing is growing in popularity for TWR testing because it enables better replication of the entire structure's field response in the laboratory than with single-axis testing. MIMO TWR can be performed using electrodynamic shakers or on a six-degree-of-freedom shaker table. Whether in an open- or closed-loop, MIMO TWR control typically involves a variant of Fourier deconvolution, wherein inputs are solved for in the frequency domain by posing the time-domain convolution with impulse response functions (IRFs) as independent multiplication problems with the frequency response functions (FRFs).

To perform TWR, accurate estimates of the system frequency response functions are essential. Extensive prior works have investigated the best practices for FRF estimation [1–3], with some focusing specifically on the TWR problem [4]. However, apparently sparsely addressed in the context of MIMO TWR are the specific "causality" issues that occur during time waveform replication. In some configurations, the standard least-squares Fourier deconvolution approach yields inputs corresponding to events in the output specification at the wrong time. For instance, if a shock event occurs at 2 seconds, Fourier deconvolution may yield high input levels at 1.9 seconds, resulting in a "pre-echo" in the output waveform, wherein the system responds to the specification before it occurs. Such causality issues are the subject of this paper. Causal system identification has been researched in some fields of electrical engineering [5–7], but many of the approaches are not applicable to TWR because they involve continuous or infinite transfer functions, or they allow extrapolation outside of the test bandwidth [8–10]. In this paper, the system FRFs are assumed to be experimentally measured and are allowed to be modified. FRF estimation for TWR is not addressed in this paper.

Keaton Coletti · Steven Carter · Ryan Schultz
Sandia National Laboratories, Albuquerque, NM 87185
e-mail: keaton.coletti@uga.edu; kpcolet@sandia.gov; spcarte@sandia.gov; rschult@sandia.gov

Brandon Dilworth et al. (eds.), *Topics in Modal Analysis & Parameter Identification II, Vol. 10*, Conference Proceedings of the Society for Experimental Mechanics Series, https://doi.org/10.13052/9788743801559-9

TWR Background

The goal of TWR is to produce a system output, $\mathbf{y}(t_n)$, $n = 1, 2, \ldots, N$, that matches a specified output, $\mathbf{y}_{\text{spec}}(t_n)$. The sample rate is given by $t_{n+1} - t_n = 1/f_s$, and $\mathbf{y}, \mathbf{y}_{\text{spec}} \in \mathbb{R}^{N_o}$. N_o is the number of output channels in the specification. The system is described by a discrete representation of the frequency response matrix (FRM), $\widehat{\mathbf{H}}(\omega_k) \in \mathbb{C}^{N_o \times N_i}$. N_i is the number of input (excitation) locations used to generate the output waveform.

Denote the discrete Fourier Transform by $\mathcal{F}$ and the inverse discrete Fourier Transform by $\mathcal{F}^{-1}$. These are typically computed using the Fast Fourier Transform (FFT) algorithm. Define $\widehat{\mathbf{y}}(\omega_k)$, $k = 1, 2, \ldots, K$ as $\mathcal{F}(\mathbf{y}(t_n))$ TWR is usually performed by solving a series of frequency-domain (FFT-transformed) least-squares problems,

$$\widehat{\mathbf{x}}(\omega_k) = \arg\min_{\mathbf{z}} \left|\left| \widehat{\mathbf{y}}_{\text{spec}}(\omega_k) - \widehat{\mathbf{H}}(\omega_k)\mathbf{z} \right|\right|_2, \tag{1}$$

where $\circ^H$ denotes the Hermitian operator, and $\hat{\mathbf{x}}$ is the frequency-domain system input. Similar formulations exist for regularized versions of the inverse problem, which consider the size of $\widehat{\mathbf{x}}$ in addition to the residuals. The time-domain and frequency-domain problem forms are related by the discrete form of the circular convolution theorem,

$$\mathbf{y}(t_n) = \mathcal{F}^{-1}\left(\mathcal{F}(\mathbf{H}(t_n))\,\mathcal{F}(\mathbf{x}(t_n))\right) = \sum_{j=0}^{N-1} \mathbf{H}_p(t_j)\,\mathbf{x}_p(t_{n-j}). \tag{2}$$

Here $\circ_p$ denotes the N-periodic infinite extension of a series. Hence, the frequency-domain TWR problem solves the question: "what infinitely repeating input will generate the infinitely repeating output specification," which has a different solution than if the system starts and ends at rest. To avoid this "convolution wraparound error," the beginning and end of the specification are zero-padded [11].

Long specifications that include many output channels can result in high or prohibitive computational cost to compute the FFT. The constant overlap add method (COLA) is often used for long TWR problems. COLA involves windowing the specification, solving for each window, and concatenating the inputs into a single sequence. Though COLA can reduce memory usage, the windowing and stitching process can introduce additional leakage errors. Hence, this paper uses a single-frame TWR framework. FFTs of the specification and IFFTs of the solved input can be computed for groups of channels in serial, and $\mathbf{x}(\omega_k)$ can be computed in groups in serial. The group size is determined by the system memory, and each group can be computed in parallel. The only condition is that the system memory should be (as a rough rule of thumb) at least eight times the size of a single channel's record. On a computer with 64 GB of memory, this is 10^9 double-precision values, or just under 3 hours of data sampled at 100 kHz. Practical applications rarely require longer data records than this.

Though the impulse response function (IRF) is the inverse Fourier Tranform of the frequency response function (FRF), the IFFT of the discrete FRF is not a discrete representation of the IRF. First, note that truncation in the time-domain is discretization in the frequency domain, and discretization in the frequency domain is truncation in the time domain. Fig. 1 shows FRF windowing (truncation) in the frequency domain, and Fig. 2 shows the same windowing in the time domain. In the top row, the operations are multiplication and truncation, and in the bottom row, the operations are convolution and downsampling. FRF truncation results in an apparent "non-causal" ring at the end of the IFFT of the FRF. The IFFT of the FRF is a finite discrete representation of the bandpass-filtered IRF, which cannot go from zero to nonzero instantly.

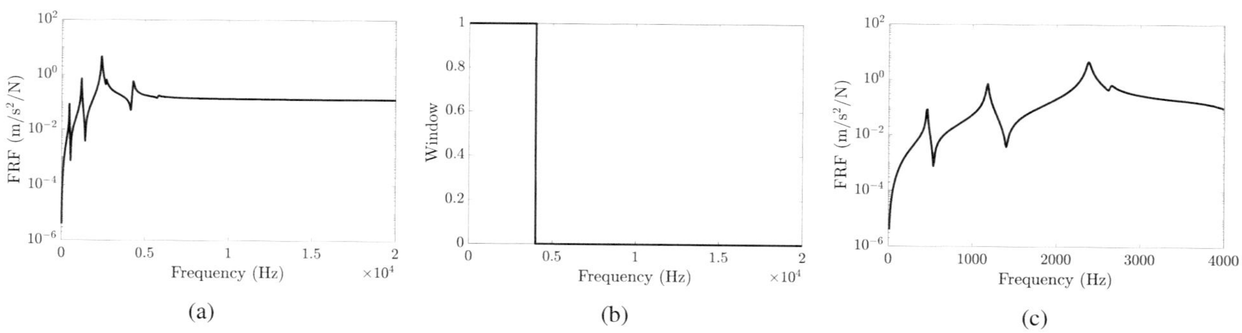

(a) (b) (c)

Fig. 1 Visualization of the effects of frequency truncation on the FRF in the frequency domain. Subfigure (a) is multiplied by (b) and truncated to yield (c).

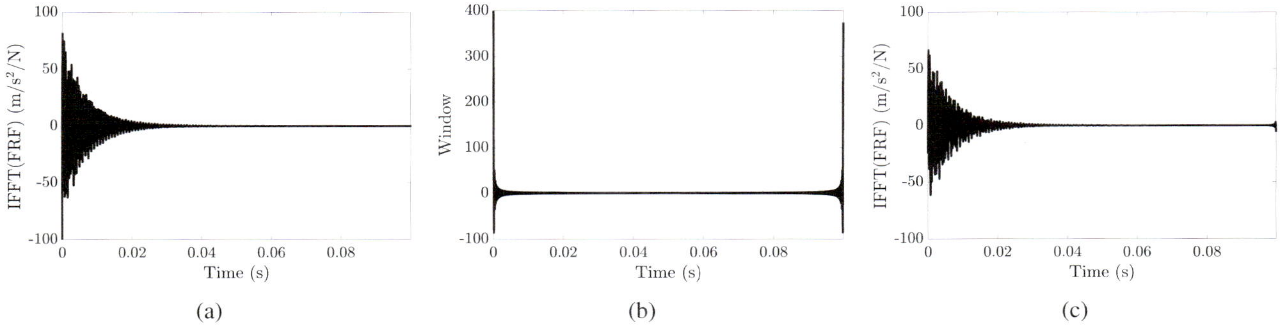

Fig. 2 Visualization of the effects of FRF frequency truncation on the estimated IRF. Subfigures are the IFFTs of the respective signals in Fig. 1. Subfigure (a) is convolved with (b) and downsampled to yield (c).

However, this bandpass artifact can be counterintuitive in light of the time-domain TWR formulation. Fig 3a shows an example of the tail of a band-limited IRF, and Fig. 3b shows the resulting input when the TWR problem is solved in a single-input, single-output configuration. The IRF tail results in input to the system before the first event in the spec, which will clearly generate a system repsponse. To visualize the analog case, the input, FRF, and specification are zero-padded in the frequency domain, which is equivalent to upsampling in the time domain. Figure 3c shows that, though the system responds before the first event in the spec, the output at the sampled time steps perfectly matches the specification. Hence, the early input to the system does not generate errors at the sampled time steps. The upsampled specification provides a reasonable visual of what the analog system output (or at least the analog specification) might look like, though it is not always intuitive. See Fig. 3d.

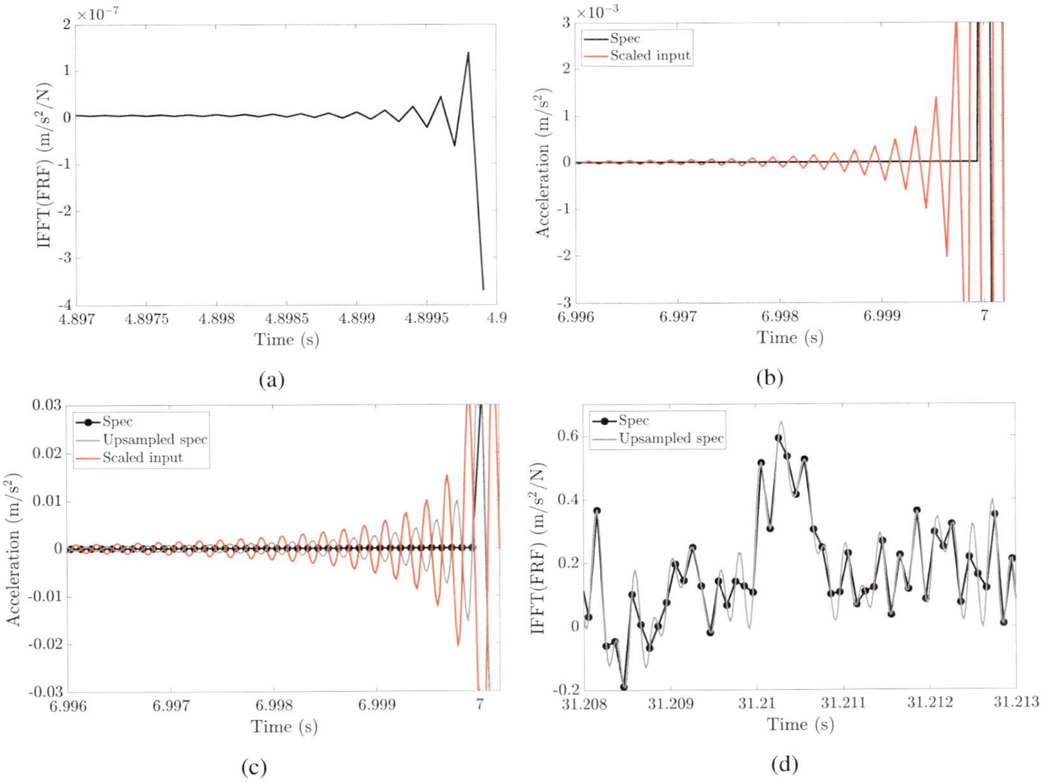

Fig. 3 Demonstration of the apparent non-causal inputs resulting from the application of a band-limited FRF to TWR. Subfigures show (a) the IFFT of a band-limited FRF (b) the first event in an ouptut specification overlayed with the solved input (c) the input, output, and specification after zero-padding in the frequency domain (d) an excerpt from the specification after zero-padding in the frequency domain.

The apparent non-causal IRF behavior caused by frequency truncation is, however, distinct from the artifacts caused by FRF measurement error. Fig. 4 shows the right tail of the IFFT of a discrete FRF with added Gaussian noise. Indeed, the tail has significant error, several orders of magnitude larger than the truncation effects in Fig. 3. The IRF errors are reflected by a large system input prior to the first event in the specification, when the system should not be responding. This error persists throughout the duration of the specification, often resulting in an audible echo that occurs in the output before events are prescribed by the specification.

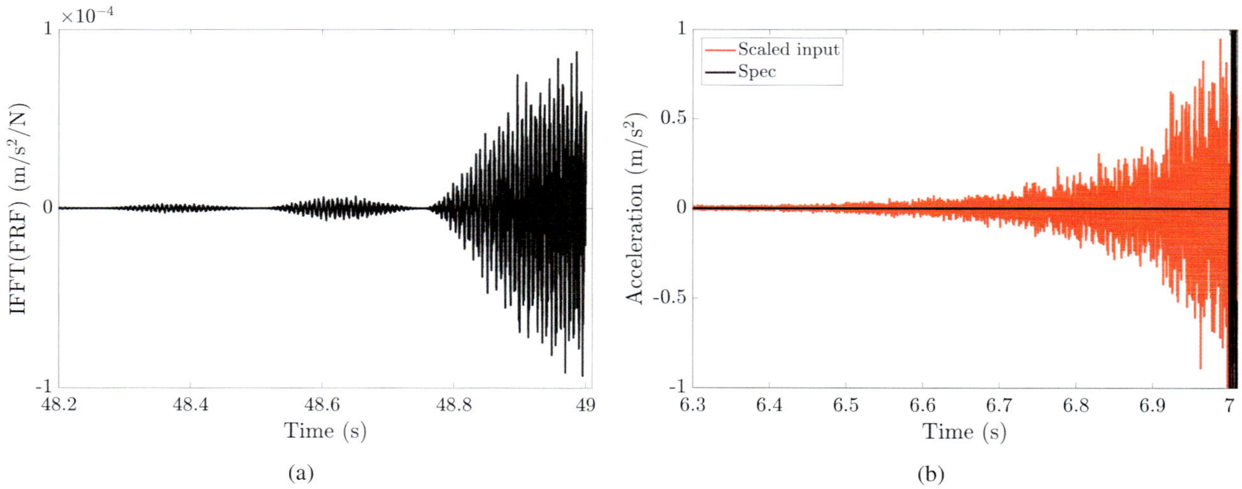

Fig. 4 Example of (a) an IFFT of a noisy, band-limited FRF and (b) the resulting non-causal system input (prior to the first event in the output specification) when that FRF is used in a TWR problem.

FRF Error Mitigation Methods

The frequency axis of $\widehat{\mathbf{H}}$ available from an experiment does not always match the frequency axis of the FFT of the specification. In the COLA framework, window sizes can be set to match the frequency resolution of the FRF. As an alternative, recent works [1, 12, 3] have proposed local modeling methods for FRF estimation. Local modeling methods rely on continuous FRF models, rather than discrete frequency bin averages, and any desired frequency axis can be specified in the FRF estimator. This work assumes that an FRF estimate has already been obtained with a frequency axis that does not necessarily match the specification. With the exception of the "minimum energy multilinear fit" method, frequency axis matching is performed in this paper using a standard complex linear interpolation. The modification methods are applied after interpolation.

Exponential taper (exp-taper) A simple, intuitive approach to correcting non-causal inputs that result from FRF noise is to remove the "ring-up" from the end of the IFFT of the FRF by multiplying it by an exponential window. Formally, the exponential taper used in this paper is

$$\widehat{\mathbf{H}}_{\text{exp-taper}}(\omega_k) = \mathcal{F}\left(\mathbf{H}(t_n)\, w(t_n)\right),$$
$$w(t_n) = \begin{cases} 1, & n \le N - N_w \\ \exp\left[-\frac{5}{N_w(n+N_w-N+6)}\right], & n > N - N_w. \end{cases} \tag{3}$$

N_w is the number of points in the IFFT that are modified, and N is the total number of points. Though applying the exponential taper prevents the input signal from depending on future specification data (Fig. 4), it changes the system FRF, generating new errors. The effect of applying the exponential taper with $N_w = .05N$ is shown in the time and frequency domains in Fig. 5.

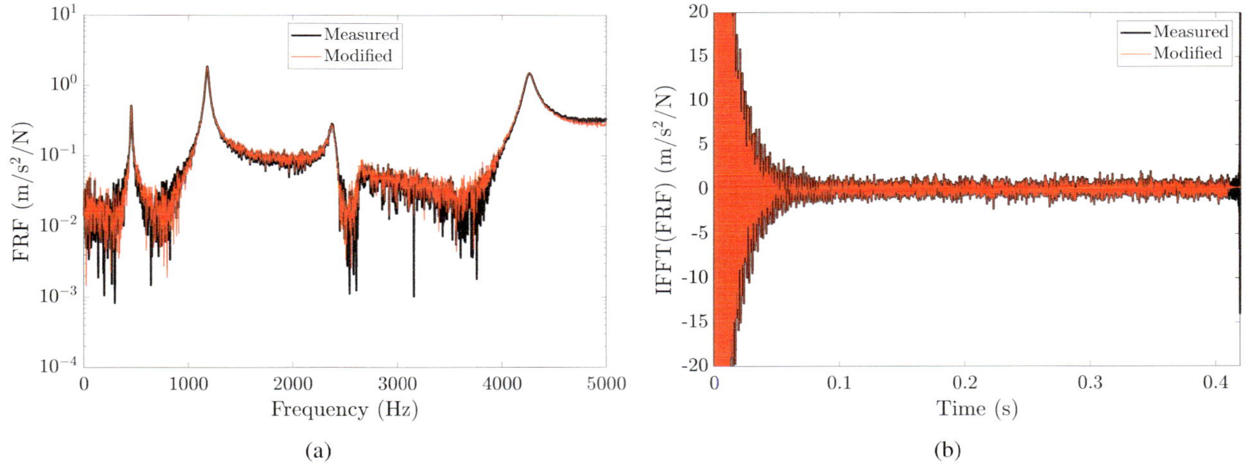

Fig. 5 Effects of the exponential taper modification method on (a) a simulated FRF (b) the corresponding band-limited IRF.

Zero-pad, truncate (pad-trunc) The exponential taper removes the bandpass-filter end effects in the experimental IRF. Hence, applying a the exponential taper to a larger bandwidth FRF could reduce the effects, since a very high-frequency FRF should have negligible bandpass filter artifacts. The "pad-trunc" method is straightforward: zero-pad the FRF, apply the standard exponential taper (IFFT $\rightarrow$ taper $\rightarrow$ FFT), and truncate the FRF. If the alteration effects are largely confined to the higher (truncated frequencies), causality problems may be mitigated with a smaller effect on the bandwidth of interest. In this paper, the FRFs are zero-padded to ten times the original frequency band. An example of the pad-trunc modification method is shown in Fig. 6. A ring-up is still visible at the end of the IFFT, as desired, but the measured FRF is significantly altered in the bandwidth of interest, expecially around the anti-resonance at 3 kHz.

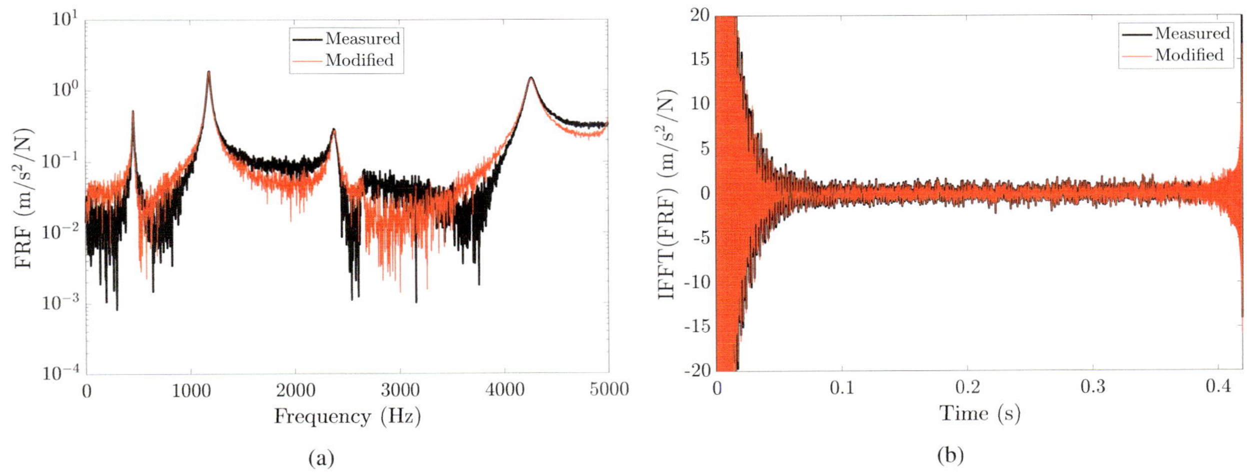

Fig. 6 Effects of the zero pad, truncate modification method on (a) a simulated FRF (b) the corresponding band-limited IRF.

Optimal Hilbert causality enforcement (hilbert) In continuous linear systems, a causal system is defined as one whose impulse response function is zero for all $t < 0$. Let $\mathcal{H}$ denote the continuous hilbert transform operator. For an impulse response function $h(t)$, an equivalent definition of causality is $\mathcal{H}(h(t)) = -i\, h(t)$ [6]. The discrete Hilbert transform is a linear mapping, so applying the discrete version of the Hilbert transform identity is a simple linear constraint. Define $\mathbf{h}$ to be a vector containing the entries in the FRF $h(t_n)$ and let $\mathbf{Ah}$ contain the entries in $\mathcal{H}(h(t_n))$, the discrete Hilbert transform.

Further, let $\mathbf{Bh} = \mathcal{F}(\mathbf{h})$. From a constrained optimization perspective, a reasonable objective is to compute

$$\min_{\mathbf{h}} \left(\mathbf{Bh} - \widehat{\mathbf{h}}_{\text{prior}}\right)^H \boldsymbol{\Sigma}^{-1} \left(\mathbf{Bh} - \widehat{\mathbf{h}}_{\text{prior}}\right) \\ \text{s.t.}\ (\mathbf{A} + i\mathbf{I})\,\mathbf{h} = \mathbf{0}, \tag{4}$$

where $\widehat{\mathbf{h}}_{\text{prior}}$ is the experimental FRF estimate. The constraint is always feasible, and a solution can be found by taking $\mathbf{h} = \mathbf{Nv}$, where $\mathbf{N}$ is the nullspace of $\mathbf{A} + i\mathbf{I}$. The optimimum is given by $\mathbf{N}^+(\boldsymbol{\Sigma})\,\widehat{\mathbf{x}}_{\text{prior}}$, where $\mathbf{N}^+(\boldsymbol{\Sigma})$ is the generalized pseudo-inverse. This method can be computationally prohibitive in large problems because $\mathbf{A}$ is $K \times K$. However, experience shows that if $\boldsymbol{\Sigma} = \mathbf{I}$, the Hilbert-optimal modification is very similar to setting $\mathbf{H}(t_n) = \mathbf{0}$ when $n = 0$ or $n \geq \lceil N/2 \rceil$. For the simulations in this paper, this form is used to compute the "hilbert" modification. Fig. 7 shows an example.

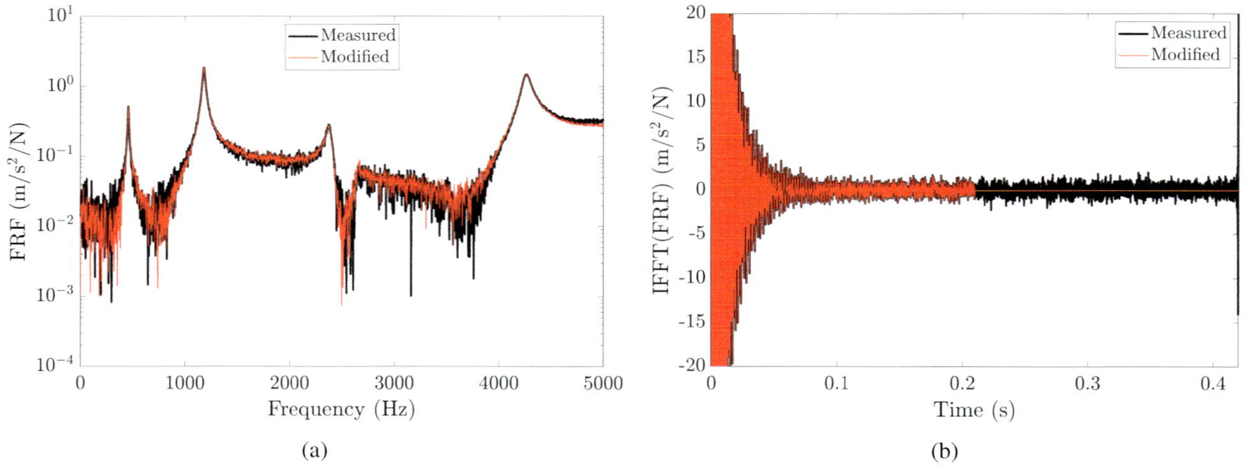

Fig. 7 Effects of the Hilbert Transform contraint modification method on (a) a simulated FRF (b) the corresponding band-limited IRF.

Minimum energy multilinear fit (linear) Basis functions for FRF smoothing procedures are typically polynomial or rational functions. An alternative method, proposed herein, is to use single-mode FRFs as a basis for smoothing. Model j is given as a function of the frequency vector, $\boldsymbol{\omega}$, by

$$\mathbf{m}_j = \frac{1}{-\boldsymbol{\omega}^2 + 2i\zeta_j \boldsymbol{\omega}\omega_j + \omega_j^2}. \tag{5}$$

To reconstruct the de-noised FRF, $\mathbf{m}_j$ are assembled into the columns of $\mathbf{M}$, and weights of each linear model are selected according to

$$\min_{\mathbf{v}} \sigma_n \left(\mathbf{Mv} - \widehat{\mathbf{h}}_{\text{prior}}\right)^H \left(\mathbf{Mv} - \widehat{\mathbf{h}}_{\text{prior}}\right) + \frac{\sigma_v}{2} \mathbf{v}^T \boldsymbol{\Sigma}_v^{-1} \mathbf{v}. \tag{6}$$

Here σ_n and σ_v represent the energy of the noise and source vector, respectively. For a vibratory system, the weights of modal FRFs are determined by the mode shapes and should be strictly real, so the real and imaginary parts are separated to form a real problem,

$$\min_{\mathbf{v}} \sigma_n \left(\overline{\mathbf{M}}\mathbf{v} - \overline{\mathbf{h}}\right)^T \left(\overline{\mathbf{M}}\mathbf{v} - \overline{\mathbf{h}}\right) + \sigma_v \mathbf{v}^T \boldsymbol{\Sigma}_v^{-1} \mathbf{v}, \\ \overline{\mathbf{M}} = \begin{bmatrix} \text{Real}(\mathbf{M}) \\ -\text{Imag}(\mathbf{M}) \end{bmatrix}, \overline{\mathbf{h}} = \begin{bmatrix} \text{Real}\left(\widehat{\mathbf{h}}_{\text{prior}}\right) \\ \text{Imag}\left(\widehat{\mathbf{h}}_{\text{prior}}\right) \end{bmatrix}. \tag{7}$$

This is a Tikhonov-regularized inverse solution to the least-squares problem $\mathbf{Mv} = \mathbf{x}$, and the selection of the parameters σ_n and σ_v is performed automatically using the procedure described in refs [13, 14]. Once the parameters are selected, the regularized inverse is given by

$$\hat{\mathbf{v}} = \left(\overline{\mathbf{M}}^T \overline{\mathbf{M}} + \frac{\sigma_n}{\sigma_v} \boldsymbol{\Sigma}_v^{-1}\right)^{-1} \overline{\mathbf{M}}^T \overline{\mathbf{h}}. \tag{8}$$

The covariance matrix, $\mathbf{\Sigma}_v$, represents the penalty on the size of each entry in $\mathbf{v}$. Setting $\mathbf{\Sigma}_v = \mathbf{I}$ risks overfitting, especially because low-frequency, lightly dampled modes have very narrow resonance peaks. If a sparse experimental FRF is curve-fit, low-frequency noise can result in large, lightly damped modes where measurement noise dominates. To ensure that the Tikhonov regularization works as desired and minimizing the energy of the curve fit, the magnitude of each curve $\mathbf{m}_j$ is squared and integrated over frequency to yield p_j, the power associated with curve j. The covariance is specified as a diagonal matrix with $\mathbf{\Sigma}_{v,jj} = p_j^{-1}$ so that the penalty term carries an equal weight for the contribution of each curve to the total energy of the fitted FRF.

Curve fitting is performed using a basis of 1200 single-degree-of-freedom FRFs with varying resonance frequencies and damping ratios.. Frequencies ω_j are set my multiplying the 40 most prominent peaks in the measured FRF by $\{0.99, 1.00, 1.01\}$ and combining every resulting frequency with modal dampings $\{0.005, 0.01, \ldots, 0.05\}$. The set of curves can be tailored to the particular application. This process is distinct from modal parameter selection: all of the curves are allowed to contribute to the smoothed FRF (even curves with overlapping frequencies), so the resulting FRF is composed of linear FRFs but is not a modal model. The results of the multilinear fit applied to the simulation in Fig. 9 are show in Fig. 8.

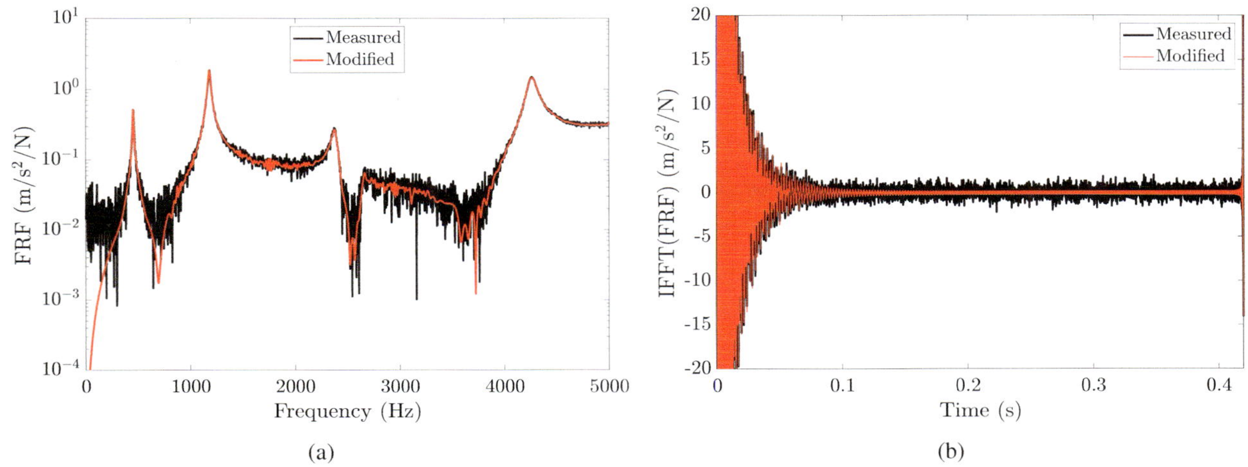

Fig. 8 Effects of the minimum-energy multilinear fit on (a) a simulated FRF (b) the corresponding band-limited IRF.

Simulation Results

The simulation is a simple linear spring-mass assembly with eight degrees of freedom (DOFs). Modal damping is set to 1% for all modes, and the FRF is truncated at 5 kHz as shown in Fig. 9b. Complex Gaussian noise is added at a specified signal-to-noise ratio. The specification is derived from an excerpt from a rock music song that is bandpass-filtered to a maximum frequency of 5 kHz and resampled at 10 kHz. The two-channel audio is distributed across the eight DOFs as shown in Fig. 9a.

Table 1 Masses and stiffnesses in the simulation structure.

	$j=1$	$j=2$	$j=3$	$j=4$	$j=5$	$j=6$	$j=7$	$j=8$
m_j (kg)	8	4	5.5	6	2	1	10	6
k_j (10^9 N/m)	2	1	2	4	2	0.5	5	1

Fig. 10 shows performance of the modification methods across a range of signal-to-noise ratios (SNRs). The error metric used is a normalized sum of squared errors,

$$\text{error} = \frac{\sum_n \left[\mathbf{y}\left(t_n\right) - \mathbf{y}_{\text{spec}}\left(t_n\right)\right]^2}{\sum_n \left[\mathbf{y}_{\text{spec}}\left(t_n\right)\right]^2}. \tag{9}$$

Each data point shows the mean of ten random trials with identical parameters. The linear fit method performs the best by a significant margin. However, in a simple linear simulation, this is not surprising, so the method should be receive

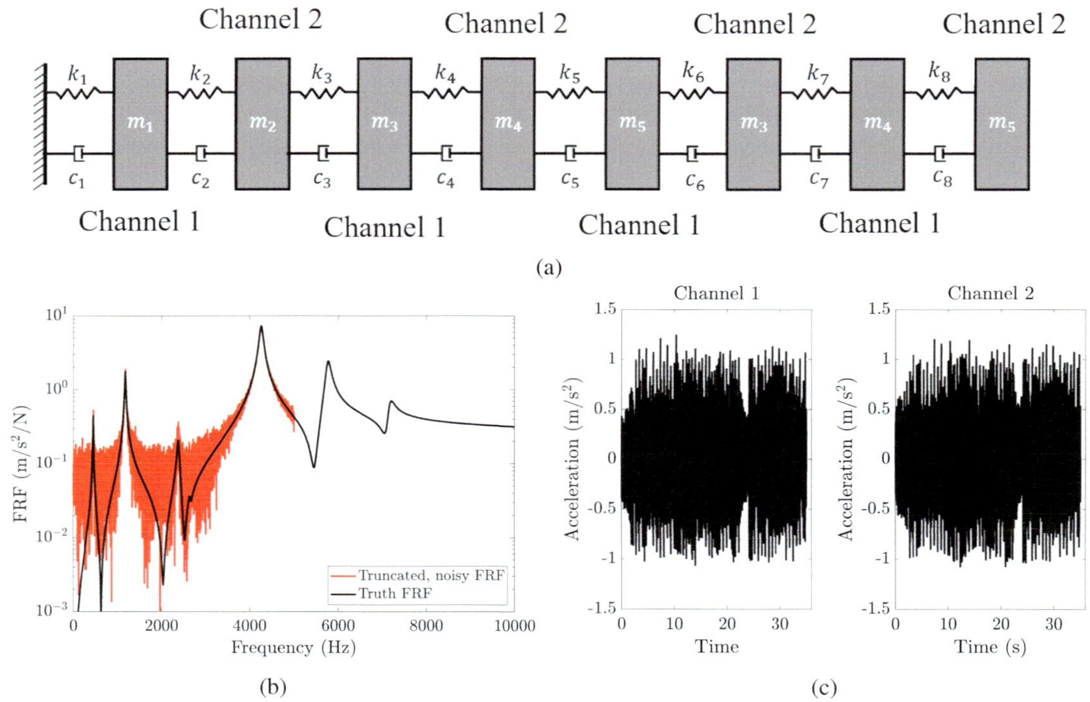

Fig. 9 Simulation case used for verification. Subfigures are (a) the simulation structure with values given in Table 1, (b) a noisy FRF (25 dB SNR), and (c) the two-channel output waveform specification.

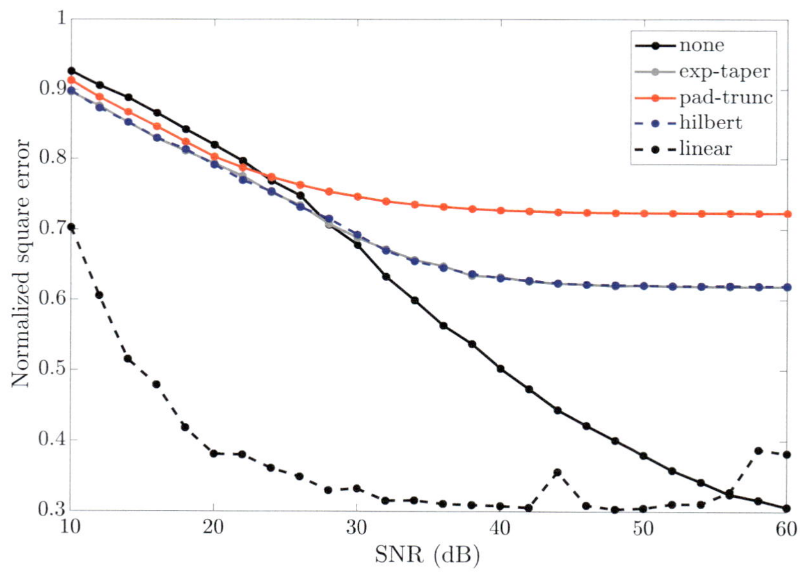

Fig. 10 Brief comparison of TWR performance for the four discussed FRF modification methods across a range of signal-to-noise ratios (SNRs).

further testing and validation. In low-noise scenarios (about 30 dB and higher SNR), applying the exponential taper or Hilbert transform modification increases error. That is, the bias generated by FRF modification exceeds the benefits of noise mitigation. In noisy scenarios, the exponential taper and Hilber modifications remove more error than they generate, yielding modest improvements in TWR.

Conclusion

This paper provides an overview of open-loop time waveform replication and describes problems associated with apparent non-causal impulse response functions. Errors in TWR are shown to result from FRF error and not from a frequency truncation in the measured FRF. In a window-free TWR framework, several FRF noise mitigation techniques are discussed and compared. In a MIMO simulation, simple FRF modifications are shown to improve TWR in very noisy cases, but they increase error in low-noise configurations. A minimum-energy multilinear FRF smoother reduces error significantly at most SNRs.

This paper provides a framework for TWR causality investigation, but many questions remain. Gaussian error was tested, and results may change if there are systematic FRF errors (such as those from leakage) or if the system is nonlinear. Further, theoretical analysis is needed to precisely determine when applying an FRF modification is benefitial and when it is detrimental. Finally, an experimental study should be performed, in particular to validate the "linear" method, which shows promise in a basic linear simulation.

Acknowledgments Sandia National Laboratories is a multi-mission laboratory managed and operated by National Technology & Engineering Solutions of Sandia, LLC (NTESS), a wholly owned subsidiary of Honeywell International Inc., for the U.S. Department of Energy's National Nuclear Security Administration (DOE/NNSA) under contract DE-NA0003525. This written work is authored by an employee of NTESS. The employee, not NTESS, owns the right, title and interest in and to the written work and is responsible for its contents. Any subjective views or opinions that might be expressed in the written work do not necessarily represent the views of the U.S. Government. The publisher acknowledges that the U.S. Government retains a non-exclusive, paid-up, irrevocable, world-wide license to publish or reproduce the published form of this written work or allow others to do so, for U.S. Government purposes. The DOE will provide public access to results of federally sponsored research in accordance with the DOE Public Access Plan.

References

1. Robbert Voorhoeve, Annemiek van der Maas, and Tom Oomen. Non-parametric identification of multivariable systems: A local rational modeling approach with application to a vibration isolation benchmark. *Mechanical Systems and Signal Processing*, 105:129–152, 2018.
2. Randall J. Allemang, Rohit S. Patwardhan, Murali M. Kolluri, and Allyn W. Phillips. Frequency response function estimation techniques and the corresponding coherence functions: A review and update. *Mechanical Systems and Signal Processing*, 162:108100, 2022.
3. Keaton Coletti, Ryan Schultz, and Steven Carter. A practitioner's guide to local FRF estimation. Technical Report SAND2023-07734C, Albuquerque, NM, February 2024.
4. Bram Cornelis, Alessandro Toso, Wouter Verpoest, and Bart Peeters. Improved MIMO FRF estimation and model updating for robust time waveform replication on durability test rigs. In *International Conference on Noise and Vibration Engineering*, Leuven, Belgium, September 2014.
5. Rodrigo A. Gonzalez, Patricio E. Valenzuela, Cristian R. Rojas, and Ricardo A. Rojas. Optimal enforcement of causality in non-parametric transfer function estimation. *IEEE Control Systems Letters*, 1(2):268–273, 2017.
6. P. Triverio and S. Grivet-Talocia. On checking causality of bandlimited sampled frequency responses. In *2006 Ph.D. Research in Microelectronics and Electronics*, Ontranto, Italy, June 2006. IEEE.
7. L. Faes, A. Porta, and G. Nollo. Testing frequency-domain causality in multivariate time series. *IEEE Transactions on Biomedical Engineering*, 57(8):1897–1906, 2010.
8. Jaeyong Cho, Karam Hwang, Seungil Jeung, and Seungyoung Ahn. An efficient extrapolation method of band-limited S-parameters for extracting causal impulse responses. *IEEE Transactions on Computer-Aided Design of Integrated Circuits and Systems*, 38(11):2086–2098, 2019.
9. Lyudmyla L. Barannyk, Hazem A. Aboutaleb, Aicha Elshabini, and Fred D. Barlow. Spectrally accurate causality enforcement using SVD-based Fourier continuations for high-speed digital interconnects. *IEEE Transactions on Components, Packaging and Manufacturing Technology*, 5(7):991–1005, 2015.
10. Jaeyong Cho, Jangyong Ahn, Jongwook Kim, Jaehyoung Park, Yujun Shin, Kibeom Kim, Junsung Choi, and Seungyoung Ahn. Low- and high-frequency extrapolation of band-limited frequency responses to extract delay causal time responses. *IEEE Transactions on Electromagnetic Compatibility*, 63(3):888–901, 2021.
11. W. G., William H. Press, Saul A. Teukolsky, William T. Vetterling, and Brian P. Flannery. *Numerical Recipes in Fortran 77: The Art of Scientific Computing.*, volume 1 of *Fortran Numerical Recipes*. Cambridge University Press, 1986.
12. Rik Pintelton, Dries Peumans, Gerd Vandersteen, and John Lataire. Frequency response function measurements via local rational modeling, revisited. *IEEE Transactions on Instrumentation and Measurement*, 70, 2021.
13. A. Pereira, J. Antoni, and Q. Leclère. Empirical Bayesian regularization of the inverse acoustic problem. *Applied Acoustics*, 97:11–29, 2015.
14. Garrett K. Lopp and Ryan Schultz. Bayesian-based response expansion and uncertainty quantification using sparse measurement sets. *Mechanical Systems and Signal Processing*, 169:107566, 2022.